D0712820

Integrated Circuits

Integrated Circuits

A User's Handbook

Michael M. Cirovic
California Polytechnic State University
San Luis Obispo, California

RESTON PUBLISHING COMPANY, INC.
Reston, Virginia
A Prentice-Hall Company

Library of Congress Cataloging in Publication Data

Cirovic, Michael M
   Integrated circuits.

   Includes index.
   1. Integrated circuits. I. Title.
TK 7874.C55   621.381'73   77-9022
ISBN 0-87909-356-0

© 1977 by
Reston Publishing Company, Inc.
A Prentice-Hall Company
Reston, Virginia 22090

10 9 8 7 6 5 4 3 2

Printed in the United States of America.

Dedicated to my wife

CAROLE

in gratitude for her patience and understanding

Contents

Preface

Integrated circuits are so readily and inexpensively available that they have changed everyone's life to date and certainly will have a significant impact in the future. This book is aimed at providing a practical introduction to the varied and numerous types of ICs. While a single volume could not possibly describe every IC available, this book attempts to provide the reader with an understanding of a large segment of the ICs that are available and with an insight into using ICs. IC manufacturers publish (non-periodically) data books and application notes. The reader is encouraged to obtain as many of these as possible; this book does not attempt in any way to compete with data books. Perhaps the clearest distinction between data books and this one is that, while in this treatment specific applications are discussed, the emphasis is on understanding the internal operation of ICs. This distinction is significant: if the IC user does not fully understand the operation of a given IC, he is at the mercy of the application engineers. He can only use the IC in the rather limited number of applications specified by the manufacturer. Furthermore, the user will not be capable of making the all-important decision as to which IC to use in a given application.

On the other hand, once the internal operation of the IC is understood, not only can the choice of IC be made intelligently, but also a world of new applications for any given IC is opened: the data sheets provided by the manufacturer make sense in specifying the characteristics of the IC, and only the ingenuity and imagination of the user limit the applications for any IC. Almost without exception, ICs can be used for so much more than the manufacturers can list in their applications literature that the user is seriously hampered if he does not take the trouble to understand the operation of ICs.

Two recent developments in ICs are not treated here: microprocessors and

I²L. Microprocessors are basically small-scale computers on a chip. As such, they offer tremendous capability at an unbelievably low cost. (The predecessors to microprocessors are the chips used in hand-held electronic calculators.) This extremely important topic is not treated here for a number of reasons: first, there are so many different manufacturers producing different microprocessors (each not interchangeable with another); and secondly, the manufacturers themselves recognize the need for more than a plain data sheet so they publish user's manuals, complete books covering a single microprocessor and its peripherals. In the case of I²L (also called injection or current logic), the development is more a technological one than one on par with T²L or CMOS, where complete logic families are available. The importance of I²L is that it offers the advantage of bipolar devices (high speed) and MOS devices (high device density). As such, I²L will never make individual gates, counters, etc., available in separate chips; rather the advantage of I²L is that it can be used to provide large-scale digital processing systems on a single chip. The trend at present is to make both the input and the output of I²L chips TTL compatible, thus making I²L chips a special case of TTL.

Section 1 of this book serves as the general introduction to integrated circuits, including the circuit design "tricks" indigenous to integrated circuits.

The second section deals with the ubiquitous analog building-block—the operational amplifier. Conventional bipolar OP AMPs, as well as special purpose and multiple OP AMP ICs, are treated in detail.

Voltage regulator ICs are covered in the third section. TTL and CMOS logic families are treated in Sections 4 and 5, respectively.

Under the heading of interface ICs in Section 6, a number of ICs important in making signals from one type of IC compatible with other ICs are discussed.

Timing ICs, VCOs, and DVM ICs comprise Section 7. Lastly, the Appendix contains a selection of data sheets for fast and easy reference.

The author is indebted to the design and application engineers from National Semiconductor, Siliconix and Signetics for their full and eager cooperation in providing information as well as samples for testing.

Lastly, the author wishes to express his gratitude for the numerous questions posed by his Senior Project students; the frequency, nature and degree of questions regarding ICs served as the guide for this book.

Michael M. Cirovic

1

Introduction to Integrated Circuits

In the field of electronics, there have been a number of "revolutionary" developments that have in some cases literally changed the course of history. The age of electronics started inauspiciously enough with the advent of vacuum tubes. The first "revolution" came in the fifties with the advent and wide-spread use of the first solid-state device—the transistor. In fact, the word *transistor* (compressing the words *transfer resistor*) took on another meaning when people started calling their new miniature portable radios transistors.

The second "revolution" was the integrated circuit (IC), first produced and utilized in the early sixties. Just as transistors offered reduced size, lower power consumption and higher reliability than their predecessors the vacuum tubes, so the ICs offered reduced size and improved reliability. Just as transistors became common in everyday life, making high-speed computers possible; the IC has taken transistors one step further. The higher circuit complexity possible in ICs brought us to the 21st century in the seventies. Within the integrated circuit field are a large number of technologies, each of which began a mild "revolution" at the time, for each new development added to our ability to utilize the possibilities contained within electronics more fully.

However, the third major "revolution" cannot be pinpointed to a precise moment in history. This "revolution" has occurred over a long period of time, and will probably continue for many years. It has been in progress without any fanfare or mention on the evening news; nevertheless, it is perhaps the most important one in terms of its significance for the average person, as well as for those in the field of electronics. This revolution is one of *cost*! For unless transistors and integrated circuits can be manufactured cheaply enough, their significance is nil to all except the curious rich. In fact, not only has the manufacturing cost of transistors and especially ICs been reduced (as with any new item, manufacturing experience and volume production usually reduce costs), but also the quality of the products has improved as well.

While inflation has pushed up the cost of almost everything in the late sixties and early seventies, the cost of ICs has been steadily coming down. This is truly significant, since a low enough cost for the IC means that the finished product, be it a digital wristwatch, a pocket computer or a lifesaving pacemaker, can be made inexpensively enough to be within the reach of almost all consumers.

The purpose of this book is to introduce the reader to a wide variety of integrated circuits. The book is written to be of most value to the users of ICs—engineers, technicians, students or hobbyists. The attempt has been made to provide a level of understanding of the circuit techniques used in ICs as well as the terminal characteristics of ICs. This approach is the only one possible, because there are so many ICs available that no single book could contain information on all. However, the understanding to be gained through the representative sample of ICs discussed here can be used to good advantage even when dealing with ICs not discussed in this book. Note that a working knowledge of basic circuit analysis and transistors (bipolar and field-effect) is assumed.

1.1   ADVANTAGES OF ICs

An integrated circuit is a single functional block which contains many individual devices (transistors, resistors, capacitors, etc.). The first and perhaps most obvious advantage of ICs is their size. The working part of a transistor is quite small, but since people are to use it, it is packaged with leads attached; thus the size is of necessity relatively large. In an IC, many interconnections between transistors and other components have been made internally, with only those terminals that are necessary being made available externally.

Another obvious advantage, closely related to size, is the drastically lower weight of ICs when compared to discrete versions. This is especially evident in large systems (like computers and airborne electronics), where decreased bulk and weight are extremely important.

Integrated circuits also offer higher reliability, simply because any given function can be implemented with fewer components.

Probably one of the most important advantages of ICs is the high level of circuit complexity made available in a small package. Because of this feature, the user of ICs can contemplate and relatively easily implement complex systems. For example, a pocket calculator with slide rule functions is a practical impossibility using discrete devices (transistors, resistors, etc.), not only because of physical size but also due to the enormously high parts count that would be required. (This would make the cost of parts as well as assembly so high as to make the unit unsalable.) The high complexity offered on a single chip has other advantages—it opens up to the user a much wider range of projects and products that can be made operational with a much smaller investment of money and manpower. A person using ICs can implement a system in a few weeks or months—which would have been nearly impossible in a lifetime, using discrete devices. In addition, the power of ICs may be reflected in the improved operation of a given system. For example, in many applications a regulated power supply is not essential, but would improve system performance. Therefore, if we had to design a voltage regulator using discrete devices the decision would be to forego the regulator in the given system. However, since available IC voltage regulators are low cost and easy to use, we would probably decide to include the regulator and thus enhance the system's performance. Another example of this kind of impact by ICs is in TV receivers. With essentially no increase in cost, TV manufacturers have incorporated additional (nonessential) features like automatic fine tuning, automatic hue (color), etc., through the use of ICs.

Figure 1.1 shows a representative sample of different integrated circuits and their packages. It should be pointed out that the figure by no means includes all possible IC configurations, just some of the more common ones.

FIGURE 1.1   AN ASSORTMENT OF ICs

1.2   BASICS OF DEVICE FABRICATION

The purpose of this section is to provide information on fabrication techniques which explains the characteristics of devices in integrated form, rather than to discuss the methods of fabricating ICs.

Perhaps the most significant single difference between discrete components and integrated circuits is the factor governing costs. As a general rule, in discrete circuit design the fewer the components used, the cheaper the circuit. This is usually not the case in integrated circuits, as cost is mainly dictated by the chip area required, and does not increase proportionally with the number of devices. This is because the devices in an IC are formed simultaneously, that is, all transistors (say *NPN*) are formed in the same step, all the collectors are made together, all the bases, etc.

Therefore, if an improvement in an IC's performance can be achieved through the use of an additional transistor, it can be incorporated with little or no increase in cost. (Since the area required for the additional device is extremely small, the increase in cost is essentially due to any change in yield, i.e. the percentage of fabricated ICs that are usable; the yield is obviously governed by the complexity of the circuit.) However, in a discrete version, the same transistor might not be incorporated unless the improvement in performance was significant enough to offset the accompanying increase in cost.

In general, the IC begins as a slab of silicon only a few mils thick and 2 to 3 inches in diameter (this is called a wafer). The processing is done under controlled conditions to prevent contamination. A number of masks are made to photographically expose the surface of the wafer on which an oxide layer has been formed and a photosensitive solution has been deposited. The wafer is then processed further to remove the oxide at selected spots. At this time, dopants can be diffused into the exposed substrate, or additional layers of silicon can be grown onto the wafer (this latter process is called epitaxial growth). This whole process is repeated as many times as necessary, with the final step of depositing a thin layer of aluminum over the entire IC. The aluminum is then selectively etched to leave the desired conductor pattern interconnecting the proper devices on the IC, as well as to provide connecting pads which will later be wirebonded to the external leads prior to packaging.

Note that during the process of forming one IC on the wafer (a single IC may require approximately 50 by 50 mils in area depending on the complexity of the function being implemented), the mask pattern is repeated so that literally hundreds of ICs are being processed simultaneously. The wafer containing many ICs is then scribed and the individual ICs are then separated, tested, connected to headers and encapsulated in the appropriate package.

Many ICs are made on a *P*-type substrate. The schematic representation of an IC *NPN* transistor is shown in Fig. 1.2. The diagram is not to scale, and does not include the higher conductivity areas near the metal contacts which ensure that they be non-rectifying or ohmic. A key feature that needs to be recognized is this: if the mask contains identical windows for the fabrication of two or more transistors, because they are made at the same time, on the same substrate and under identical conditions, these transistors will possess almost identical terminal characteristics. Thus, *matched transistors can be readily fabricated in an IC*. Furthermore, unbalancing the masks for two transistors by making the emitter area of one larger by a certain factor will have the effect of making the β of that transistor larger by the same factor. Thus, although it is not practical (nor easily done) to try and provide a specific, accurately controlled value of β, *the ratio of β's for transistors on an IC can be readily controlled*.

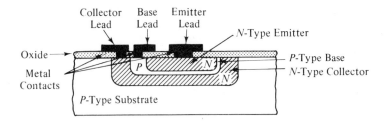

FIGURE 1.2 INTEGRATED CIRCUIT *NPN* TRANSISTOR (CROSS-SECTIONAL VIEW)

Two variations of the basic *NPN* IC transistor commonly used are the *multiple collector transistor* and the *multiple emitter transistor*. The schematic representation for a two-collector transistor is shown in Fig. 1.3. This scheme is used to stabilize and reduce the transistor β and g_m (as discussed in the next section). The terminal characteristics are that of a transistor with a lower β and g_m than that of the conventional transistor in Fig. 1.2. The idea of more than one

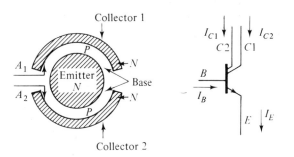

FIGURE 1.3 TWO-COLLECTOR TRANSISTOR (TOP VIEW IN CROSS-SECTION) AND CIRCUIT SYMBOL

collector can be extended to three or more collectors, as shown in Fig. 1.4. If we define β_1, β_2 and β_3 to be the ratio of I_{C_1}, I_{C_2} and I_{C_3} to I_B, this configuration yields β's which are scaled, one with respect to the other, according to the ratio of the effective collector areas. Since the collector areas are controlled through the masks and the photographic process, the end result is a transistor with multiple collectors and respective β's whose ratio is highly predictable. These properties are used to great advantage in circuit design for ICs.

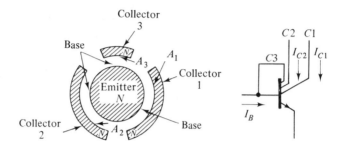

FIGURE 1.4   MULTIPLE COLLECTOR TRANSISTOR

The schematic diagram of the construction of a multiple emitter transistor is shown in Fig. 1.5. Although only two emitters are shown, the scheme can be and is extended to as many as needed. The terminal characteristics for this type of transistor are again quite different from the conventional *NPN* transistor: a much lower effective β due to the reduced emitter efficiency of the geometry and the increased effective base width. The multiple emitter configuration is mainly used in TTL circuits, although it is sometimes used in linear applications.

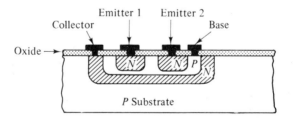

FIGURE   1.5   MULTIPLE   EMITTER   TRANSISTOR   (CROSS-SECTIONAL VIEW)

One of the factors that does control the cost of manufacturing ICs is the number of diffusion steps. Therefore, in order to minimize the number of steps,

PNP transistors are made at the same time as their *NPN* counterparts. This dictates a totally different geometry for the *PNP* transistor. The *PNP* base is diffused at the same time as the *NPN* collectors, while both the emitter and collector of the *PNP* are diffused together with the *NPN* emitter. The resulting structure, shown in Fig. 1.6, forces transistor action (injection of carriers from the emitter into the base and eventually into the collector) to occur parallel to the chip surface, and thus is named *lateral PNP transistor*. A number of terminal characteristics are affected by this unusual geometry. First, the transistor β is quite low, due to the relatively wide base and high doping of the collector. Secondly, because of the geometry and the high collector conductivity, the collector-to-base capacitance is quite high. The full significance of this will be discussed in the section on operational amplifier frequency response.

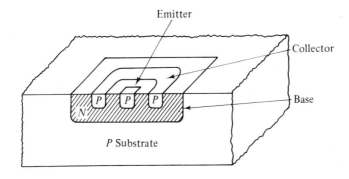

FIGURE 1.6   LATERAL *PNP* TRANSISTOR

Resistor construction is illustrated in Fig. 1.7: the collector diffusion is used to provide isolation from other devices in the IC, while the base diffusion forms the actual resistor. Since all resistors are made simultaneously, all have the same depth of diffusion, h. Therefore, the ratio of resistivity to depth, called the *sheet*

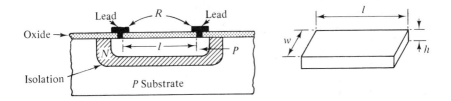

FIGURE 1.7   DIFFUSED RESISTOR

resistance, is identical for all resistors. The specific resistance is then given by:

$$R = \frac{\rho l}{h w_1} = R_S \frac{l}{w}$$

where R_S is the sheet resistance. The production control of sheet resistance is not precise, giving the resistor an absolute value which is not precisely predictable. However, note that since the ratio of any two resistors is not a function of sheet resistance:

$$\frac{R_1}{R_2} = \frac{l_1}{w_1} \cdot \frac{w_2}{l_2}$$

and since the width and length are photographically controlled through the masks, *the ratio of resistors can be controlled precisely*. This means, for example, that if we need two 10 k resistors, when they are integrated we may not have exactly 10 k resistors, but the two will have the same value.

One problem that diffused resistors have is the relatively large chip area required for large valued resistors. Thus large valued resistors are either avoided through circuit design, or *pinch resistors* are used.

Integrated capacitors are of the MOS type: one electrode is formed by the N-type silicon, the dielectric being the silicon-oxide layer, and the other electrode is provided by the metalization, as shown in Fig. 1.8. Only low values of capacitance can be attained in a reasonable amount of space, so larger values of capacitance are avoided through circuit design, or are added to the external circuit utilizing the IC.

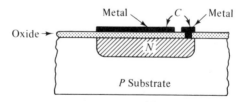

FIGURE 1.8   MOS CAPACITOR

Integrated inductance efforts have been fruitless; thus inductors are not utilized in ICs. (An inductor can, however, be synthesized using a voltage follower, two resistors and a capacitor—only for relatively low frequencies.)

A regular diode is fabricated as a transistor (*NPN*), with the collector and base shorted together during the metalization stage. The essential advantage of this method is that, in addition to not requiring any additional steps in manufacture, it gives us diodes which match the base-emitter characteristics of *NPN* transistors. Again, this fact is important in circuit design.

Similarly, an integrated zener diode is made as an *NPN* transistor with the base and collector shorted during metalization. The reverse characteristics of the base-emitter junction are utilized, giving a zener voltage of approximately 7 V.

1.3   CIRCUIT CONFIGURATIONS GENERIC TO ICs

Circuit design for ICs makes use of the special properties of devices that IC fabrication offers, while at the same time obviating some of the shortcomings. In addition, the design is flexible enough (at least in linear ICs) to allow the ICs to be used over a wide range of power supply voltages.

The basic idea of *constant-current sources* is illustrated in Figs. 1.9 and 1.10. Essentially, in the case of both the source and the sink, the voltage across the resistor R is a constant at $V - V_{BE}$. (V_{BE} stays essentially constant over a wide range of collector currents.) Thus the current through R must be constant. This is the emitter current, and the sink or source current is

$$I = \frac{\beta}{\beta + 1} I_E \cong I_E = \frac{V - V_{BE}}{R}$$

Thus if β is large (which is usually the case), the sink or source current is essentially constant for a constant V.

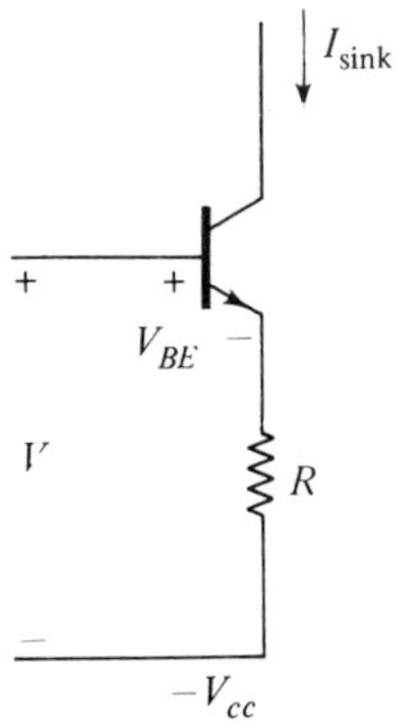

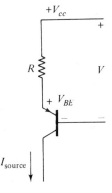

FIGURE 1.9   CONSTANT-CURRENT SINK

FIGURE 1.10   CONSTANT-CURRENT SOURCE

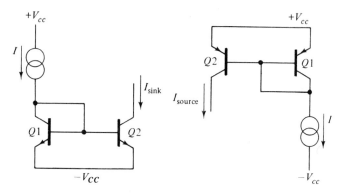

FIGURE 1.11   CURRENT MIRROR CONFIGURATIONS

Another configuration widely used in linear ICs, the *current mirror*, is shown in Fig. 1.11. Here the matching between the two transistors $Q1$ and $Q2$ is required. Given a current I into $Q1$ (note that $Q1$ is really connected as a diode), the two collector currents must be the same since their base-to-emitter voltages are identical. For a large β, where base currents may be neglected as compared with collector currents, this circuit configuration effectively gives I_{sink} (or I_{source}) equal to I. Thus the current I has been "mirrored" into I_{sink}. (Actually, $I_{sink} = I - 2I_B$. This error may be reduced with the addition of $Q3$, as shown in Fig. 1.12. For this circuit, the error between I and I_{sink} is $2I_B/(\beta + 1)$.) The current mirror circuit can be used to scale the current I if the two transistors are made with unequal emitter areas, or by the insertion of emitter resistors as shown in Fig. 1.13. Here the exact relationship between the currents is not simple; however, if we assume β very large and a negligible imbalance between the two base-emitter voltages,* then the voltage drops across R_1 and R_2 are essentially the same. Under these conditions, the reflected current is:

$$I_{sink} = \frac{R_1}{R_2}I$$

*When the voltage drops across R_1 and R_2 are small, for example below one volt, a more exact relationship between the resistor and current ratios is:

$$\frac{26mV}{I_1R_1} \; 1_n(I_2/I_1) = 1 - (I_2R_2/I_1R_1)$$

This equation may be solved graphically, or by iterative means. The Wilson circuit shown in Fig. 1.12c is a means for making the current mirrored as nearly equal to the reference current as possible and can be used in any of the current mirror configurations depicted.

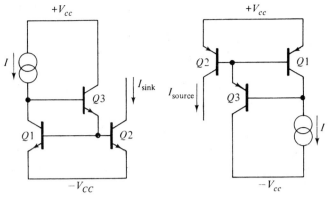

(a)   (b)

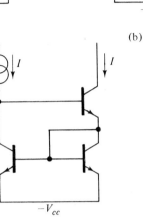

(c)

FIGURE 1.12   IMPROVED CURRENT MIRROR CIRCUIT: (a) SINK; (b) SOURCE; (c) WILSON CURRENT MIRROR

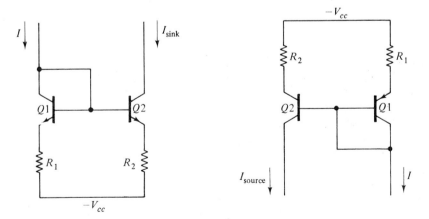

FIGURE 1.13   UNEQUAL CURRENT MIRROR CIRCUIT

11

Since in an IC the ratio of two resistors can be controlled to a relatively tight tolerance, the current I is reflected by a constant (which may be larger or smaller than 1).

When it is necessary to generate a high ac gain, the amplifier calls for a large collector resistor. To save space, an *active load* (Fig. 1.14) is commonly used rather than a resistor. If I is a constant current, then the active transistor $Q3$ is biased by the current mirror of $Q1 - Q2$. At the same time, $Q2$ acts as a load for $Q3$ which is amplifying the signal. $Q3$, operating under constant base current, offers a high impedance ($1/h_{oe}$), thus providing high voltage gain for the signal through $Q3$.

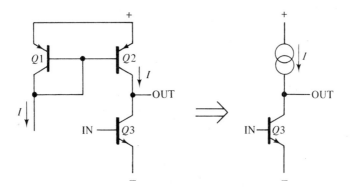

FIGURE 1.14   ACTIVE LOAD

Another common configuration that is used for level translation is shown in Fig. 1.15. If the transistor base current is neglected when compared to the current I, and assuming V_{BE} to be essentially constant over a range of collector currents, then $I = V_{BE}/R_2 =$ constant. The voltage V is then:

$$V = n V_{BE} \qquad \text{where } n = 1 + (R_1/R_2)$$

Again note that the constant voltage V is controlled by a ratio of resistors (which can be precisely set), thus providing a programmable dc drop. Any variation in the net current through the circuit causes a negligibly small change in the voltage. So this circuit functions like a programmable zener. Other methods of achieving dc level translation involve the use of zeners (reverse-biased base-emitter junctions) and a number of forward-biased diodes connected in series. The circuit of Fig. 1.15 offers the advantage of being more flexible since n can be any number larger than 1, not necessarily an integer.

The preceding discussion should serve to illustrate the distinctive features of circuit design for ICs. One thing should be plain: in an IC amplifier containing

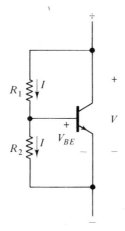

FIGURE 1.15 nV_{BE} CIRCUIT

FIGURE 1.15 nV_{BE} CIRCUIT

many transistors, only relatively few transistors act as amplifiers. In addition, the reader should be able to glance at a schematic of an IC and recognize certain circuit configurations, thus gaining a better understanding of the internal functioning of the IC. Although this understanding is not always essential, there are many applications where establishing the proper functioning of the IC does require at least some knowledge of its internal operation.

1.4   FROM SCHEMATIC TO FUNCTIONING CIRCUIT

In most cases, the uninitiated IC user follows the application notes supplied by the manufacturer to the letter (or at least that is what he thinks), and usually the circuit functions as expected. However, this is not always the case. By their very nature, manufacturers' data sheets, and in most cases application notes, cannot contain all the information necessary for a successful completion of a project or system using ICs. Besides, most ICs offer far wider capabilities and applications than just those listed in data sheets and application notes.

In this section, some of the general guidelines for the successful application of ICs to any project are outlined and discussed. More information as it pertains to specific ICs is discussed in the sections dealing with those ICs.

There are only three basic requirements for the proper use of ICs in any project. These are: *proper circuit design, proper construction and test procedures, and proper power supply choice.* As with any rules, these three are more easily stated than followed.

PROPER CIRCUIT DESIGN

This is perhaps the hardest of the three, since it requires the most from us:

1. Know and understand the properties and limitations of as large a variety of ICs as possible. Be on the lookout for new ICs, for they may meet your needs more readily than any other.
2. Use the proper IC for the purpose at hand. Don't try to make do with another circuit—in most cases it doesn't pay.
3. Double and triple check your design to ascertain that it does *exactly* what is desired—no less and no more. This especially applies to digital circuits, since they usually involve complex timing and it is not uncommon to overlook a given set of (signal) conditions which gives an erroneous output.

PROPER CONSTRUCTION AND TESTING PROCEDURES

Probably the largest percentage of problems encountered by the uninitiated involves either improper construction or inappropriate testing. A little care in construction will go a long way, be it a breadboard or a printed circuit board. First, let us consider the prototype or breadboard—the stage at which the basic circuit design is tested and evaluated. There are five methods available for constructing the prototype: the custom pc (printed circuit) board; the vectorboard with sockets and point-to-point wiring; special sockets; wirewrap sockets; and lastly, standard pc boards with a foil pattern to accommodate only the ICs (or their sockets) where interconnections are wired in to suit a particular situation. Each of these methods offers certain advantages, while also presenting limitations and its own unique problems.

No single method is best for all cases. Although the custom pc board offers the neatest and fastest construction, once the board is made, it lacks flexibility for circuit changes are not easily implemented.

The wirewrap technique is applicable where few discrete components are used and where the exact circuit configuration has not been finalized in the design stage, i.e. where wiring changes are a high probability. At the same time, the wirewrap technique is not advised where signals of 1 MHz or higher are used. This technique, although offering flexibility, makes the prototype somewhat hard to troubleshoot in terms of finding a wiring mistake or in tracing signals. (This is due to the fact that even in relatively simple circuits, the wirewrapped prototype board has many criss-crossed wires and tends to resemble a bird's nest.) This technique is especially well suited for digital circuits of great complexity.

Standard pc boards offer a good compromise between the custom pc board and point-to-point wired vectorboard. They are available in two types: one to accommodate DIPs (dual in-line package ICs), one to accommodate the round TO-5 type packages. The advantage of this scheme over the wirewrap is that it does not require special tools, and like the wirewrap technique, this scheme is probably better suited to digital applications.

Special sockets which can accommodate up to about 100 IC pins (not ICs) are a good choice for simple digital as well as analog prototypes. They provide interconnection points for wires or discrete components next to the IC pins. This scheme offers the easiest and most convenient means for changing the circuit since no soldering is involved. However, due to the high capacitance between sockets, this method is not advised for signal frequencies much above 500 kHz.

The vectorboard approach offers the widest latitude in parts location, since this is entirely determined at the user's discretion. It is probably the most tedious and slowest method—each socket and external component must be individually mounted and interconnected. However, with proper parts layout and ground-plane, this type of prototype parallels the custom pc board in terms of its frequency-handling and ability to accommodate all types of circuits—analog and digital.

In the final analysis the choice is an individual one; the best choice can be made after one has been exposed to all the different schemes. Listed below are some suggested guidelines for both the prototype as well as the final pc board (obviously not all suggestions apply to all construction schemes).

1. *Prevent and/or eliminate unwanted feedback.* Remember that both electric and magnetic fields act at a distance. So, if you do not desire a part of the output signal at the input, do not run any wires (or pc foils) from the output near the input. Also, try to position components and interconnecting wires of one part of the circuit away from other parts of the circuit.

2. Arrange the components so as to maintain a signal flow through them in as nearly a straight line as possible. This should eliminate unwanted feedback, and improve testing and troubleshooting.

3. Keep leads and wires as short as possible. Not only does this make a neater prototype, but it should aid in troubleshooting and minimize unwanted feedback. (For troubleshooting purposes it is also helpful to use wires of as many insulation colors as possible, rather than of a single color.)

4. Use sockets for ICs, at least in the prototype. An IC with, for example, 14 leads soldered into a pc board is very hard to remove without either damaging the IC or destroying the pc foil. For ease of troubleshooting as well as IC replacement, sockets are advised.

5. Break up complex circuits into smaller functional blocks and provide (in prototype) for easy separation and reconnection of these blocks. This applies to both analog and digital circuits. When a complex prototype is completed

and power is applied, if the proper output or operation is not attained, the smaller functional blocks can be.isolated and thus the trouble found more readily.

6. When soldering ICs, use a fine point *low* (25 to 30) wattage soldering iron. Remember, although the leads that come out of the IC package are hefty, inside them are wirebonds to the chip of very small diameter gold wire, and excessive heat can cause these bonds to come apart (then a perfectly good chip must be discarded because a connection has been lost).

7. When inserting ICs into sockets, make certain that the pins are aligned with the sockets—do not force the IC into the socket. Avoid bending the IC leads on DIPs and on TO-5 packages (other than to form the leads of the round package to fit a DIP socket). When using sockets, do not insert any ICs until all the wiring is finished.

8. Align all the ICs so that pin 1 (or the tab on the TO-5 package) points in the same direction for all the ICs. This will prevent inserting any of the ICs backward. (Note that item 1 in this list should take precedence.)

9. Apply power only after all the ICs have been inserted. Similarly turn off power before removing any ICs.

The proper testing of a prototype is essential. The following list should provide the basic guidelines:

1. Have a test procedure throught out and outlined *prior* to turning power on. This will ensure that the proper and necessary equipment is on hand and that all necessary tests have been performed.

2. Record all test results *along* with the serial number (or any identifying number) of the equipment used. Note that this number is *not* the model number, but a number that identifies the specific piece of equipment. This is essential, since in many cases test results themselves are not obviously correct or incorrect, but may require additional calculations or comparisons. If a reading should turn out to be questionable, having the serial number for the specific instrument used can settle the question of whether the erroneous reading was due to a malfunction in the test instrument or in the prototype.

3. Before applying power for the first time, determine one or two critical points in the circuit that can be monitored to indicate if proper operation has been achieved. If proper operation is not evident, turn off the power to avoid possible damage to any components. The most likely causes of trouble are wiring error, followed by missing wiring, incorrect component (i.e. resistor color code read incorrectly, etc.), and faulty or partly faulty components.

4.   Know the characteristics of your test instruments: input impedance, band-
     width, accuracy, etc. For example, in checking the frequency of an rf
     amplifier, make certain that the voltmeter used for measuring the output has
     sufficient bandwidth to adequately do the job. Similarly, in using an oscillo-
     scope to check a digital circuit, make certain that it has sufficient bandwidth
     to allow you to see the pulses and use a compensated probe.

5.   Make certain that the test instrument will not load down the circuit under
     test. For example, some frequency counters have a 50 Ohm input impe-
     dance, which would load down the output of most linear circuits.

6.   Make certain that you do not short out any part of the circuit with the ground
     of the test instrument.

7.   It is usually best to avoid taking readings at high impedance points in the
     circuit (i.e. OP AMP inputs, etc.).

The foregoing list is by no means all inclusive. Use it as a basis and extend it
to suit the situation.

PROPER POWER SUPPLY

This is the last consideration because it is recommended to use a good,
commercial regulated power supply for the initial prototype evaluation. In this
way the power supply is eliminated from the possible list of problems that may
originally arise. In addition, there is no need to guess or estimate the power
supply characteristics necessary for the finalized project. If a bench supply is
used, once the design has been evaluated and finalized, the current drain can be
determined exactly, and the power supply designed accordingly. The following
power supply characteristics need to be determined:

1.   The power rating—i.e. the steady-state current and voltage(s) required.

2.   The net regulation required—this is the net variation in the dc voltage that
     can be tolerated without circuit malfunction.

3.   Any special needs for power supply bypassing or decoupling. Bypassing, if
     required, should be done at the pins of the IC's power supply input. Bypass-
     ing the power supply leads a few inches from the IC is of little value. A
     ceramic disc capacitor with leads cut as short as possible should be used,
     with value typically between 0.01 and 0.1 μF. When decoupling is neces-
     sary (in high frequency circuits), a ground plane should be used. (Both sides
     of the pc board should have as much copper left on them as practical, and the
     common or ground of the supply should be connected to the excess

copper—called the *ground plane*.) In addition, the scheme shown in Fig. 1.16 should be used. This illustrates how any ac signals that may be on the supply line (be they due to switching transients or other stray signals) will appear across the series resistor and not get to the IC supply terminal. The capacitor to ground further shunts these stray signals, thus it should be ceramic and chosen just large enough to have very low impedance at the frequencies of signals involved. The resistor chosen should be as large as possible without causing more than 0.1 V drop in the dc supplied to the IC. If bipolar supplies are used, the scheme should be repeated for both supplies.

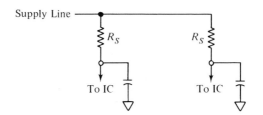

FIGURE 1.16   SUPPLY DECOUPLING

Another important aspect of power supply hookup is the problem of ground loops. We shall not deal with it here, but will consider it as necessary in dealing with specific ICs and in applications where this problem is critical.

2

Operational Amplifiers

Of all ICs, the operational amplifier (OP AMP) probably best exemplifies the versatility and capability of monolithic techniques. To analog computer users, OP AMPs are nothing new, since vacuum-tube OP AMPs have been utilized in analog computer work for many years. After the advent of transistors, discrete transistor OP AMPs replaced vacuum-tube versions in some applications. However, the OP AMP became an important circuit design device only after it was integrated and available in a small and inexpensive chip. With improvements in fabrication techniques, as well as large volume production, the cost of monolithic IC OP AMPs has dropped drastically. In fact, some high (relatively) performance OP AMPs can be purchased for less than the cost of many individual transistors or FETs.

Most simply, an OP AMP is a high gain, dc-coupled differential amplifier. In an ideal sense, it possesses:

a. Infinite voltage gain
b. Infinite input impedance
c. Zero output impedance
d. Infinite bandwidth
e. Zero noise
f. Zero parameter drift with time and temperature

Modern monolithic OP AMPs come relatively close to matching the ideal characteristics in the first three cases, while they do not perform quite so well on the last three.

Before examining terminal characteristics of OP AMPs, we shall first take a brief look at the internal circuit, since all terminal (external) characteristics are the result of the internal circuit configurations.

2.1   INTERNAL OPERATION

There are a number of complexities to be overcome in discussing the internal operation of OP AMPs. First, there are many different types available—from all bipolar, JFET and bipolar, all MOSFET, to hybrid ones containing specially selected (or trimmed) devices along with the basic monolithic circuit. It is therefore impossible to discuss all possible circuit configurations.

In addition, the same type number OP AMP manufactured by two different companies does not necessarily contain the exact same circuit. The situation is further compounded by manufacturers' occasionally updating the design of the circuit (and usually neglecting to update the schematic in the data sheets), so that the same type OP AMP, purchased from the same manufacturer one year later, may not have the same circuit as its predecessor.

Nevertheless, we can learn a great deal about the operation of monolithic OP AMPs through the example of the type 741 OP AMP, shown in Fig. 2.1. First, the obvious: the circuit requires two power supply voltages, one positive and one negative with respect to ground (or common). This allows for dc coupling of signals at both input and output with a permissible voltage swing above and below ground.

Second, there are two input terminals (the bases of $Q1$ and $Q2$), providing for differential amplification as well as basic inverting and non-inverting gain configurations.

Bias for all amplifying transistors is derived from the current through the 39 k resistor (in the middle of the circuit). Thus the current through both $Q11$ and $Q12$ is the net supply voltage minus two diode drops (2×0.6 V) divided by 39 k. This current is mirrored (see Chapter 1) and scaled down by $Q10$. The same current flows through $Q9$; this current is again mirrored by $Q8$ to provide the essentially constant current bias for the differential amplifier formed by $Q1$ and $Q2$. Transistors $Q5$ and $Q6$ form a current mirror to equally split the net current from $Q8$, so as to bias the differential pair with essentially equal currents. The differential input signal is amplified by $Q1$ and $Q2$, followed by $Q3$ and $Q4$ (note that $Q3$ and $Q4$ have active loads formed by $Q5$ and $Q6$). The signal is then further processed by the Darlington configuration of $Q16$, $Q17$. For negative output voltages, drive is applied directly from $Q17$ to the *PNP* output transistor $Q20$. For positive output voltages, the signal path from the collector of $Q17$ is

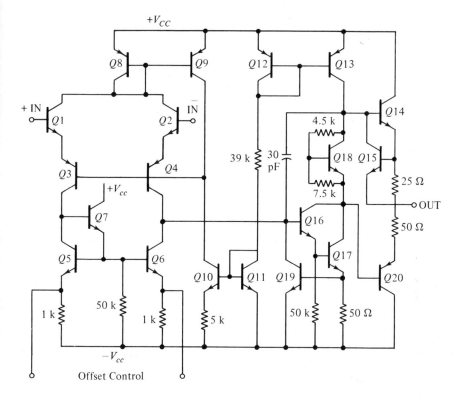

FIGURE 2.1   TYPE 741 OP AMP CIRCUIT DIAGRAM

through the dc-level shifting stage ($Q18$), to the *NPN* output transistor $Q14$. Bias for the output stage is obtained by mirroring the current through $Q12$ to $Q13$. The nV_{BE} circuit of $Q18$ provides for approximately 1 Vdc difference between the bases of $Q14$ and $Q20$, thus assuring class AB operation of the output (this reduces distortion of the output signal). The purpose of $Q15$ is to act as a current limiter for $Q14$, while $Q19$ acts in the same capacity for $Q16$, $Q17$ and therefore $Q20$.

Other versions of this circuit have a full complementary output stage with similar current-limiting as shown in Fig. 2.2. The operation of the circuit is as follows: neither $Q2$ nor $Q3$ (in Fig. 2.2) normally conducts—the output current is derived through $Q1$ (if V_o is positive) or through $Q4$ (if V_o is negative).

For positive output voltages, $Q4$ is essentially off and the output current is supplied by $Q1$, as shown in Fig. 2.3a. The output current does not normally cause a sufficient drop across R to cause $Q2$ to turn on and conduct. Thus all of the drive current I_i goes as base current for $Q1$. However, when I_o increases to a

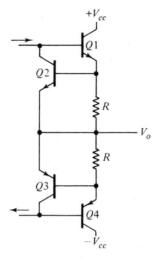

FIGURE 2.2 TYPICAL OUTPUT STAGE WITH
CURRENT-LIMITING

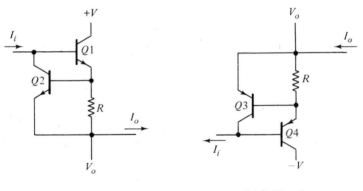

(a) Sourcing Current   (b) Sinking Current

FIGURE 2.3 OUTPUT CURRENT-LIMITING

value of V_{BE}/R, the base-emitter junction of $Q2$ is forward biased by the voltage drop across R, and $Q2$ conducts. This means that only a portion of the drive current reaches the base of $Q1$; the rest is siphoned off by the collector of $Q2$. Since the output current is roughly β times the base current of $Q1$ and I_{B_1} no longer tracks I_i, the output is in its current-limit mode. Although the drive current increases to try to maintain the output voltage under increasing load demands (load resistance decreasing), $Q2$ maintains essentially a fixed voltage across R, thus limiting the output current to V_{BE}/R. A totally symmetrical situation exists for negative output voltages involving $Q4$ and $Q3$. When the load demand exceeds V_{BE}/R in output current, $Q3$ turns on and limits the current through $Q4$ to this value. From the circuit in Fig. 2.1 we can deduce that the maximum current sourced by the 741 should be approximately 25 mA (0.6/25). The negative current limit is not that easily established, but for the symmetrical circuit it is also about 25 mA.

Some common variations on the 741 circuit are to add an additional *PNP* transistor between $Q17$ and $Q20$ to increase the internal gain for negative outputs. The lateral *PNP* transistor alone does not have a sufficiently high gain to match the *NPN*. Another change would be to replace the nV_{BE} circuit of $Q18$ by two transistors with collector to base shorts, connected in series to provide the desired dc voltage drop of approximately 1 V.

For the 741 circuit shown, the bias collector current of the input pair ($Q1$ and $Q2$) is set for approximately 10 μA. So, assuming a minimum expected β for these transistors of, for example 50, gives us an expected base bias current of 200 nA. In addition, for a collector current of 10 μA, the minimum transistor h_{ie} is:

$$h_{ie} = \frac{\beta \ (0.026 \ \text{V})}{I_c} = 130 \ \text{k} \ \Omega$$

The input impedance for a differential input (between the bases of $Q1$ and $Q2$) is $2h_{ie}$. Thus we see that it is desirable to operate the input stage at a very low current level to provide as low an input base current as possible, and as high an input impedance as possible. Another factor that improves both of these parameters (for a given collector bias current) is obviously β—the higher the β, the lower the base current and the higher the input impedance. Conventional integrated *NPN* transistors typically have a β of 200. However, super-β transistors, fabricated by adding a second emitter diffusion, offer β's of 5000 and higher. Second generation OP AMPs like the 307, 318, and others utilize improved techniques like these to provide significantly higher input impedance and much lower input currents than the 741.

Basically, the OP AMP circuit contains voltage amplification and biasing stages. Each of the amplifying transistors exhibits junction capacitance at higher frequencies. This is evidenced by lower voltage gain as well as significant phase shift between the input and output. The OP AMP is mainly used with negative feedback provided externally. However, at higher frequencies, the internal phase shift would cause the negative feedback to become positive, i.e. regenerative. This would cause the amplifier to oscillate at the frequency where the internal OP AMP phase shift is 180 degrees. To prevent this situation from occurring, all OP AMPs must be compensated. Some, like the 741, are internally compensated, while others require external compensation. In the 741, the 30 pF capacitor, used between the base of $Q16$ and the collector of $Q17$, compensates for this internal phase shift by causing the gain to decrease below unity at the frequencies where the phase shift would otherwise cause the amplifier to oscillate.

Besides determining the small-signal bandwidth, this compensation capacitor, together with the current source, sets the large signal performance. For example, when the output voltage should change rapidly from a negative to a positive value, the largest current available to charge the capacitor is $2I_{c2}$ (assuming all of the current for the differential amplifier to be through $Q2$). Thus the rate of change of the capacitor voltage (and thus the output voltage) is limited to:

$$\frac{dV_o}{dt} = \frac{2I_{c2}}{C} \cong 0.7 \text{ Volts/microsecond for 741}$$

We shall return to both the small-signal bandwidth and *slew-rate* (the maximum time-rate-of-change in output voltage) in subsequent sections.

An obvious circuit modification to improve the slew-rate for the OP AMP is to simply allow for a higher current for charging the capacitor, or for smaller capacitance, or for both. In the case of smaller capacitance, the stage corresponding to $Q16$-$Q17$ in the 741 is replaced by one with a lower gain, thus allowing a smaller value of capacitance to compensate for excess phase shift. Such techniques result in OP AMPs with slew-rates as high as 500 V/μs.

Another group of OP AMPs, usually termed low power or programmable, allows for external control of the overall bias current. An example of such a type is shown in Fig. 2.4. By connecting an external resistor between the collector of $Q9$ and V_{cc}, the collector current of $Q9$ is set:

$$I_{c9} \cong \frac{2V_{cc}}{R_s}$$

where R_s is the externally connected resistor. Since the bias for all the stages is derived by mirroring of the current in $Q9$ (by transistors $Q10$, $Q11$, and $Q12$), this external resistor sets the bias for all stages. In addition, the resistor sets the net supply current. Such an OP AMP (the L144 is another example—see Appendix) is programmable to the extent that the externally adjustable resistor R_s, by setting bias currents, controls the input bias current, the total supply current, the small-signal gain (all transistor parameters are a function of the bias current), the input impedance, and the slew-rate.

In most respects the rest of the 4250 circuit is similar to the 741, except for the use of *PNP* input transistors, $Q1$ and $Q2$, the use of two diode drop bias instead of the nV_{BE} circuit in the output, and the altered configuration of the Darlington pair, $Q5$-$Q8$. The use of *PNP-NPN* Darlington is to allow operation from a more greatly reduced power supply voltage than those required by the 741.

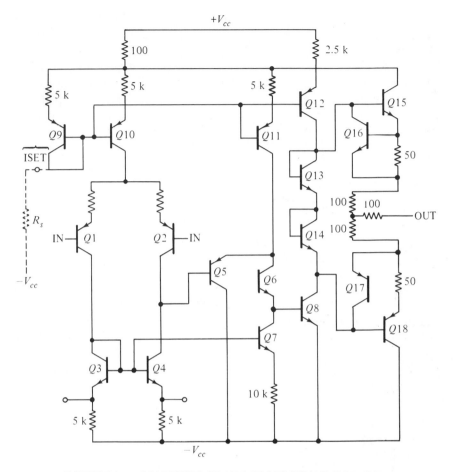

FIGURE 2.4   SCHEMATIC OF 4250 PROGRAMMABLE OP AMP

The power supply voltage required by a particular OP AMP is directly determined by the internal circuit configuration. Although all OP AMPs are designed to perform over a wide range of power supply voltages, each has a minimum as well as a maximum supply voltage. The minimum is usually determined by the amount of voltage required to provide bias for all the transistor junctions, as well as sufficient bias current for reliable operation. The maximum supply voltage is usually determined by either the reverse breakdown voltage of the collector-base junctions of the transistors, or by the individual and collective device power dissipation maximum. This illustrates yet another way in which IC design is different from discrete-component design: the power dissipated in one device affects the temperature of the whole chip, and therefore has the effect of decreasing the power that can be safely dissipated in another device. This thermal coupling between devices is utilized to advantage in voltage regulator ICs.

The foregoing discussion of the internal operation of OP AMPs, although limited in scope, should allow the user of OP AMPs to have a better understanding of the causes and controlling factors of the terminal characteristics of OP AMPs. Only through such an understanding can one make the proper choice of an OP AMP for a particular application or, conversely, get the most in performance from a particular OP AMP.

2.2   OP AMP PARAMETERS

In discussing the internal operation of OP AMPs we have already mentioned some of the terminal characteristics. The specific definitions and significance of commonly specified OP AMP parameters are listed here.

LARGE SIGNAL VOLTAGE GAIN   This is the ratio of output voltage (maximum swing) to the difference in voltage between the two input terminals required to produce the output. Note that this is the open-loop gain of the OP AMP without any feedback, so it is the largest gain possible. Any (negative) feedback produces closed loop gains lower than this value. Typically, this parameter is in excess of 50,000 (it might also be specified as 50 V/mV).

INPUT RESISTANCE   The effective resistance seen at either input terminal with the other one grounded. Again, this parameter is defined for the open-loop amplifier. Typically, this value is in excess of 1 M Ω, and may be as high as 10^{14} Ω for FET input OP AMPs.

OUTPUT RESISTANCE   The effective (small-signal) resistance that appears in series with the output (assuming that the output has been modelled by an ideal voltage generator). Typically, this value is around 100 Ω.

INPUT OFFSET VOLTAGE   The voltage by which the two input terminals must be offset in order to cause zero output voltage. It is more practically the uncompensated resulting output voltage when the OP AMP is operated as a voltage follower with the input grounded. This offset results from the mismatch in the base-emitter characteristics of the two input transistors. Typically, the input offset voltage is a few mV, and can be significantly lower for high quality OP AMPs.

INPUT BIAS CURRENT   The actual dc base current required by each of the two input transistors. (Sometimes, it is defined as the average of the two base currents.) Note that this current enters the input terminals for *NPN* input transistors, whereas for *PNP* input transistors it leaves the input. Typically, the input bias current is 200 nA, significantly less for super-β input transistor OP AMPs (less than 1 nA), and even smaller for FET input OP AMPs (less than 1 pA).

INPUT OFFSET CURRENT   The difference between the two input base currents. It is caused by the mismatch in the base-emitter characteristics of the input transistors. Typically, the input offset current is one order of magnitude smaller than the input bias current.

INPUT VOLTAGE RANGE   The maximum positive and negative voltage that may be applied to either input terminal and maintain proper bias of the input transistors. This parameter is defined for specific power supply voltages; typically, the input terminals cannot be brought closer than within one or two V_{BE} of the supply voltage.

OUTPUT VOLTAGE SWING   The maximum excursion (positive and negative) of the output voltage specified for specific power supply voltages. Like the input voltage range, the output voltage swing is limited by the need to maintain bias on the current sources and amplifying transistors internal to the OP AMP. The output voltage can typically swing to within one or two V_{BE} of either positive or negative supply voltage (depending on the specific circuit configuration).

POWER SUPPLY REJECTION RATIO   The ratio of input offset voltage to the change in power supply voltage producing the input offset voltage. As a result of a change in the supply voltage, the master bias current, and therefore all bias currents, changes; specifically the bias on the input differential pair is changed. This is reflected as an additional input offset voltage. However, the mirrored bias currents do not change in value proportionally with the master bias current; therefore the change in bias on the input stage is not as large. Furthermore, since the input is differential, changes in bias of the two differentially connected transistors tend to cancel. The power supply rejection ratio (PSRR) is typically in excess of 80 dB. It is also sometimes specified in μV/V, meaning that the stated number of μV change results from a 1 V change in supply voltage. For example, a PSRR of 100 dB is the same as one specified as 10 μV/V.

COMMON-MODE REJECTION RATIO   Usually defined as the ratio of the difference gain to the common-mode gain, and usually expressed in dB, typically in excess of 80 dB. While it is an important parameter since it gives an indication of how closely a given OP AMP resembles an ideal OP AMP, the common-mode rejection ratio (CMRR) is at the same time one of the most widely misunderstood parameters.

First, let us represent the complete OP AMP circuit by the symbol shown in Fig. 2.5, where V_1 and V_2 are connected to the bases of $Q2$ and $Q1$ respectively

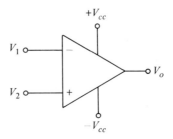

FIGURE 2.5   OP AMP SYMBOL

(Fig. 2.1). The power supply leads are shown here to emphasize that power must be applied; they will be assumed to be connected and will not be shown in subsequent diagrams.

If we short V_2 to ground, the output is strictly due to V_1—this defines the inverting gain (thus the minus sign at this input) A_1. Similarly, if an input is applied to V_2 with V_1 connected to ground, the resulting output voltage defines the non-inverting gain A_2. Thus with both inputs applied, the output voltage is:

$$V_o = A_1 V_1 + A_2 V_2$$

Because of circuit symmetry, the two gains A_1 and A_2 should be numerically almost equal, with A_1 negative and A_2 positive. The two gains are not identical due to the slight mismatch that may be present in the input differential stage. (For an ideal OP AMP $A_1 = -A_2$.) This deviation from ideal behavior, although slight, causes the output voltage to be not strictly a function of the difference in the two input voltages. As a measure of the error, consider V_1 and V_2 as represented in Fig. 2.6. The difference voltage is:

$$V_d = V_1 - V_2$$

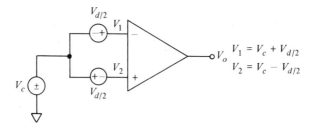

FIGURE 2.6   COMMON-MODE AND DIFFERENCE SIGNALS

The common-mode signal is the average of the two input signals referred to ground:

$$V_c = \tfrac{1}{2} (V_1 + V_2)$$

In an ideal amplifier, the output would be only a function of the difference signal and not of the common-mode signal. The connection in Fig. 2.7 allows us to measure the difference gain, A_d, defined as the ratio of output to input voltage when there is no common-mode signal ($V_c = 0$). For the voltage polarities shown, this number is negative and approximately equal to the open-loop gain.

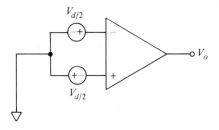

FIGURE 2.7   DIFFERENCE GAIN

The circuit of Fig. 2.8 allows us to measure the common-mode gain, A_c, defined as the ratio of output to input voltage when there is no difference signal ($V_d = 0$). This number is usually small and may be positive or negative. By linear superposition, the output voltage for neither V_d nor V_c zero is:

$$V_o = A_d V_d + A_c V_c$$

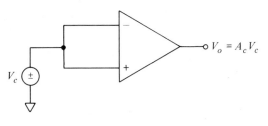

FIGURE 2.8   COMMON-MODE GAIN

The two representations shown in Figs. 2.5 and 2.6 are equivalent. It is then a relatively simple task to show that, with the difference and common-mode signals as defined, the two sets of gains are related:

$$A_d = \tfrac{1}{2}(A_1 - A_2)$$
$$A_c = A_1 + A_2$$

With A_1 negative and A_2 positive, we readily see that A_d is a large negative number, while A_c is close to zero. Whether A_c is positive or negative is determined by which of the two gains, inverting or non-inverting, is numerically larger.

To see the significance of this non-ideal operation, let us take the following example.

With $V_1 = + 0.1$ mV and $V_2 = - 0.1$ mV, the resulting output is $V_o = - 10$ V. The experiment is repeated with $V_1 = + 1000.1$ mV and $V_2 = + 999.9$ mV to get an output $V_o = - 10.5$ V. In both cases the difference input is 0.2 mV, while the common-mode input is zero in the first case and 1000 mV in the second. From the first set of conditions we can calculate the difference gain:

$$A_d = - 10/0.0002 = + 50{,}000$$

From the second set of conditions, and with A_d now known, we can calculate the common-mode gain:

$$A_c = \frac{[- 10.5 - (- 50{,}000 \times 0.0002)]}{1.0001 - 0.9999} = - 0.5$$

The common-mode rejection ratio is then:

$$\mathrm{CMRR} = \frac{A_d}{A_c} = \frac{50{,}000}{0.5} = 100{,}000 \qquad \text{or } 100 \text{ dB}$$

Note that had the output voltage for the second set of conditions been $- 9.5$ (which is just as likely), A_c would have been $+ 0.5$, but the CMRR would still have been the same as calculated above. This is because the important quantity is the amount of error in the output (in the above example, the error is 0.5 V for the second case), and not whether the error is positive or negative. Thus it is the absolute value of the ratio of the differential to the common-mode gain that is specified.

To see how the non-infinite CMRR (for an ideal amplifier $A_c = 0$ and CMRR is infinite) affects the gain, we can rewrite the output voltage in terms of the CMRR and difference gain:

$$V_o = A_d V_d \left(1 + \frac{A_c V_c}{A_d V_d}\right) = A_d V_d (1 + V_c/\mathrm{CMRR} \times V_d)$$

The last form of the output voltage equation is useful in two ways. First, for given CMRR, common-mode and difference input voltages, we can predict the error caused by using the approximate relationship:

$$V_o = A \ (V_1 - V_2)$$

where A is the difference or open-loop gain. For example, with a CMRR of 90 dB and a common-mode signal which is less than 100 times the difference signal, the approximate relationship results in less than 1% error.

Second, for a desired maximum error, with the input common-mode and difference voltages known, we can determine the minimum CMRR required. For example, for a maximum output error of 0.1%, with $V_d = 2$ mV and $V_c = 100$ mV, the CMRR required would be approximately 94 dB.

In examining the data sheets, such information is very valuable in determining the appropriate device for a particular application.

It is essential to note that in listing the OP AMP parameters, manufacturers specify the conditions under which these parameters are guaranteed. To see how different operating conditions affect the different parameters let us examine the manufacturer's data sheet.

Again we shall use the 741 OP AMP as an example. Its small-signal frequency response, magnitude, and phase are shown in Fig. 2.9. Note that the open-loop gain near dc is extremely large (in excess of 10^5 or 100 dB), and that, due to the compensating capacitor, the gain starts to fall off at the rate of 20

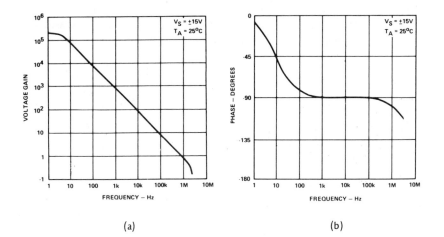

(a)   (b)

FIGURE 2.9   FREQUENCY RESPONSE (741): (a) OPEN-LOOP GAIN AS A FUNCTION OF FREQUENCY; (b) OPEN-LOOP PHASE RESPONSE AS A FUNCTION OF FREQUENCY

dB/decade until at 1 MHz the gain is unity. This value is typical of most OP
AMPs and is called the *gain-bandwidth product*.

As for discrete transistors, so it is for the OP AMP: most parameters are
sensitive to and depend on the junction temperature and power supply voltage.
Figure 2.10 shows the dependence of the input bias current on temperature. As
might be expected, at higher temperatures a lower base current is needed to cause
given collector current.

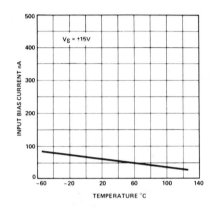

FIGURE 2.10   INPUT BIAS CURRENT AS A FUNCTION OF AMBIENT
TEMPERATURE (741)

The functional dependence of the input resistance on temperature is shown in
Fig. 2.11. Since the input resistance is a function of h_{ie}, which increases with
temperature, the input resistance, as expected, also increases with temperature.

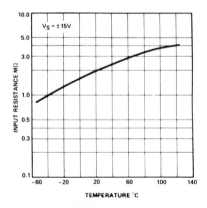

FIGURE 2.11   INPUT RESISTANCE AS A FUNCTION OF AMBIENT
TEMPERATURE (741)

In a 741, the total supply current is determined by the supply voltage, since the master bias current is proportional to the 39 k resistor (see Fig. 2.12) and all other bias currents are mirrored from the master. Therefore, the power consumption (the product of supply voltage and total supply current) also increases with increasing power supply voltage. However, since the bias currents do not increase linearly with supply voltage, the power consumption does not increase as the square of the supply voltage (as would be the case in a purely resistive circuit). The power consumption is almost linearly related to the supply voltage as indicated in Fig. 2.12.

As mentioned earlier, the maximum output voltage swing is to within a few V_{BE} of the supply voltage. This is shown in Fig. 2.13. For example, with ± 15 V supplies (net supply voltage of 30 V), the typical maximum voltage swing at the output is 25 to 26 V.

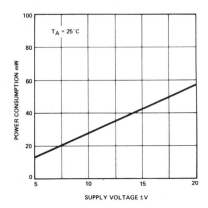

FIGURE 2.12   POWER CONSUMPTION AS A FUNCTION OF SUPPLY VOLTAGE (741)

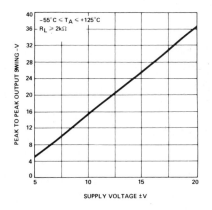

FIGURE 2.13   OUTPUT VOLTAGE SWING AS A FUNCTION OF SUP-PLY VOLTAGE (741)

The open-loop voltage gain increases with supply voltage, as shown in Fig. 2.14. The bias currents all increase with supply voltage, however not proportionally. The gain of each stage also increases with supply voltage, but not in a direct ratio. So, the open-loop gain does not increase linearly with supply voltage.

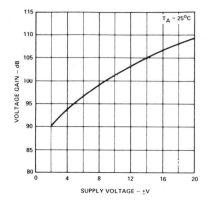

FIGURE 2.14   OPEN-LOOP VOLTAGE GAIN AS A FUNCTION OF SUPPLY VOLTAGE (741)

At higher frequencies, the mismatch in the differential input stage increases, and therefore the CMRR decreases with frequency, as shown in Fig. 2.15.

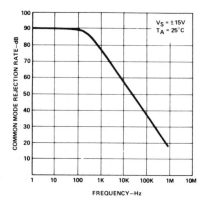

FIGURE 2.15   COMMON-MODE REJECTION RATIO AS A FUNCTION OF FREQUENCY (741)

The maximum output voltage swing is essentially constant for frequencies below 10 kHz, and falls off rapidly for frequencies above 10 kHz due to current saturation of the input stage; this is slew-rate limited. A typical plot is shown in Fig. 2.16.

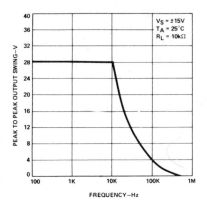

FIGURE 2.16   OUTPUT VOLTAGE SWING AS A FUNCTION OF FRE-QUENCY (741)

The small-signal transient response is shown in Fig. 2.17. The classical rise time, defined as the time to go from 10% to 90% of the final output voltage, is approximately 0.3 μs, which gives a slew-rate of approximately 100 mV/μs.

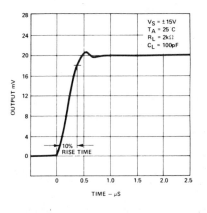

FIGURE 2.17   TRANSIENT RESPONSE (741)

2.3   BASIC AMPLIFIER CONFIGURATIONS

As we have seen, the OP AMP is a dc-coupled high-gain amplifier. In fact the open-loop gain is so high that negative feedback must be provided externally. The most basic OP AMP gain stage is shown in Fig. 2.18. If we assume the

CMRR and open-loop gains to be extremely high (even infinite), then for any finite output voltage, the voltage difference across the OP AMP inputs must be essentially zero. So, as far as voltage is concerned, the OP AMP input presents a short (zero volts). However, remember that the bias current is extremely small, so that if the currents I_1 and I_2 are in mA, then the input current may be effectively neglected. Thus I_1 must equal I_2. We then note that $I_1 = Vi/R_1$ and also $V_o = -I_2 R_2$. Substituting, we obtain the gain:

$$\frac{V_o}{V_i} = -\frac{R_2}{R_1}$$

It is obvious from the above gain expression that the gain is set by the ratio of two resistors and is essentially independent of OP AMP parameters. Anyone who has had to bias discrete transistors and try to stabilize the operating point as well as gain will readily appreciate the simplicity of making an amplifier as shown in Fig. 2.18. This is the simplest OP AMP configuration, providing inverting gain. Note that the magnitude of the gain can be less than one as well as greater than one. The upper limit is set by the OP AMP open-loop gain. Further note that the gain is not a function of the OP AMP (so long as it has a high enough open-loop gain and low enough input bias current).

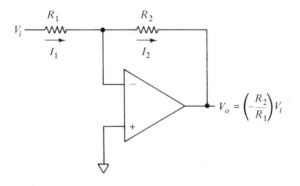

FIGURE 2.18   INVERTING AMPLIFIER

The basic premises of the foregoing analysis stand us in good stead for *all* OP AMP configurations incorporating negative feedback. We can then generalize the means of analyzing any OP AMP circuit with negative feedback: assuming the OP AMP not to be voltage, current or slew-rate limited, *the output will sink or source the current required to force the difference between the inverting and non-inverting inputs to be essentially zero, while drawing negligible current between the two input terminals*. This is a rather unique condition that the OP AMP sets up at the input: a short for voltage and an open for current. This may be termed a *virtual short*.

The non-inverting gain function is performed by the circuit configuration shown in Fig. 2.19. We can analyze this circuit using the same procedure as above. If the voltage across the input terminals is zero, then V_i must appear at the junction of R_1 and R_2. The current through R_1 is then V_i/R_1. Again, if the inverting input draws negligible current, then I_1 and I_2 must be equal. The output voltage is then the drop across both R_1 and R_2. Making the appropriate substitutions, we obtain the gain:

$$\frac{V_o}{V_i} = \frac{R_1 + R_2}{R_1}$$

Again, the gain is set by resistors, and is not a function of the OP AMP. However, in the non-inverting case, the gain is always larger than 1 for any values of resistors R_1 and R_2.

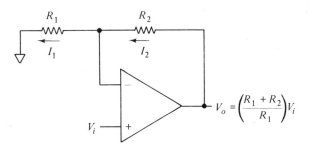

FIGURE 2.19   NON-INVERTING AMPLIFIER

A special case of the non-inverting amplifier is worth considering—the case where R_1 is infinite (open circuit) and R_2 is zero (short circuit). This is shown in Fig. 2.20. As we can readily see from the non-inverting gain equation, letting R_1 approach infinity and setting R_2 at zero, the gain for this configuration is unity.

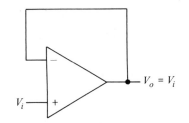

FIGURE 2.20   VOLTAGE FOLLOWER

Thus the circuit is a voltage follower, and has numerous applications in filter design as well as buffering or impedance transformation applications. The non-inverting input offers a high impedance to the source, while the output offers a low impedance to the load.

2.4   OFFSET VOLTAGE AND BIAS CURRENT COMPENSATION

Although the worst case values of input offset voltage as well as input offset current are quite low, many applications require that the errors caused by them be eliminated.

Most OP AMPs have external connections available for adjusting the current balance of the input stage to cancel the input offset voltage. The input circuit of the 741 OP AMP, repeated in Fig. 2.21, is representative of the means provided for this adjustment. In the case of a 741, a 10 k potentiometer (10-turn for critical applications) is connected to the negative supply as shown. Other OP AMPs may require 100 k or other value potentiometers connected to the positive or negative supply. Care should be taken to connect the potentiometer to the polarity called for by the manufacturer of the OP AMP used, since irreparable damage could result from connection to the wrong supply. Whatever the application, in order to trim the offset voltage the input should be shorted to ground (making certain that

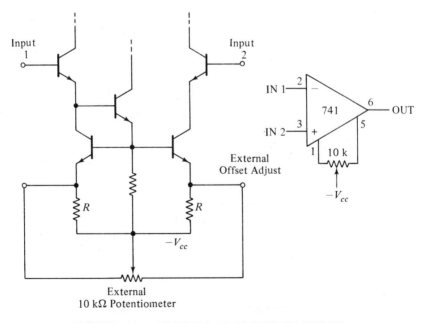

FIGURE 2.21   TRIMMING INPUT OFFSET VOLTAGE

the previous stage, if any, is disconnected first); then the potentiometer is adjusted until a sensitive voltmeter (one that can *accurately* resolve a millivolt) connected to the output reads zero. It is important to note that this adjustment is valid only for the specific OP AMP for which it has been *performed*, and that if the OP AMP is replaced or substituted (even with one of the same type), the nulling procedure must be repeated. Note also that the input offset voltage for any OP AMP can be easily measured by connecting it in the voltage follower mode with the input at ground. The output voltage is then equal to the input offset voltage.

The possible error due to input bias current is largest for very low input voltages since the input bias current is then the same order of magnitude as the signal currents. The compensation scheme shown in Fig. 2.22 offers a simple means for cancelling the error in an inverting amplifier. If the two (inverting and non-inverting) bias currents are assumed equal, with V_i zero, V_o should be zero.

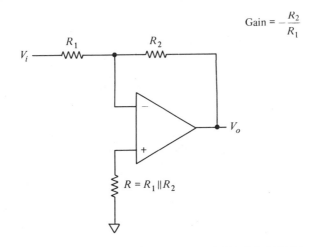

$$\text{Gain} = -\frac{R_2}{R_1}$$

FIGURE 2.22   SIMPLE BIAS CURRENT COMPENSATION

If V_o is indeed zero, then R_1 and R_2 appear in parallel from the inverting input to ground. Thus to cancel the voltage drop caused at the inverting input, we need only insert an equal resistance in series with the non-inverting input. Thus the compensating resistor R should have a value equal to the parallel combination of R_1 and R_2. This technique eliminates the error due to the input bias current (for any OP AMP), but it does not eliminate the error due to the input offset current. However, for a specific OP AMP, if R is replaced with a potentiometer (wiper connected to one end) having a value approximately 20 to 30% higher than the parallel combination of R_1 and R_2, the potentiometer can be adjusted with the input shorted to ground for zero output voltage. This will eliminate the effect of both the input bias current and the input offset current.

In the non-inverting mode, with a high source dc impedance, the configuration shown in Fig. 2.23 should be used. The value of the compensating resistor R required depends on the source resistance:

$$R = R_s \ - \ \frac{R_1 R_2}{R_1 + R_2}$$

In this manner, the resistance seen from the inverting and non-inverting inputs to ground has been equalized.

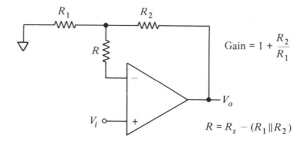

FIGURE 2.23   BIAS CURRENT COMPENSATION FOR HIGH SOURCE
IMPEDANCE

Similarly, for the non-inverting mode with low source impedance, equalizing the impedances from the inverting and non-inverting inputs to ground, is accomplished by inserting the compensating resistor in series with the input. This is shown in Fig. 2.24. The value of compensating resistance required is:

$$R = \frac{R_1 R_2}{R_1 + R_2} - R_s$$

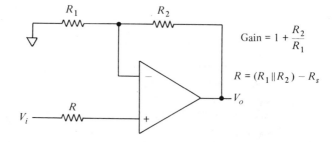

FIGURE 2.24   BIAS CURRENT COMPENSATION FOR LOW SOURCE
IMPEDANCE

It should be noted that a low source impedance is assumed for the inverting amplifier case, since, if that were not the case, a large gain error would result, obviating any need of compensation for input bias current.

More elaborate schemes for actually cancelling the input bias current for the inverting and non-inverting configurations are shown in Figs. 2.25 and 2.26, respectively. In both cases, the input bias current of the OP AMP is provided by the external *PNP* transistor's base current, thus requiring essentially no current from the input. The cancellation current is set by the resistor R, which must be

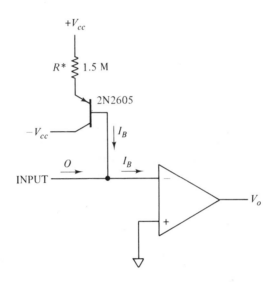

FIGURE 2.25   IMPROVED BIAS CURRENT CANCELLATION (INVERT-
ING MODE)

trimmed for each specific OP AMP. The more elaborate circuit for the non-inverting configuration is required to accommodate the fact that the non-inverting input in this case need not be at ground. In addition, the use of *PNP* external transistors is dictated by OP AMPs having *NPN* input stages; while for OP AMPs with *PNP* input stages the basic circuit is repeated with *NPN* external transistors (instead of *PNP*) connected to the negative supply instead of the positive.

For the voltage follower operating from high source impedances (almost always the case), the compensating resistor is included between the inverting input and the output, as shown in Fig. 2.27. The value of R should equal the source impedance.

In most cases, the simple compensation schemes are sufficient, while for critical applications the cancellation technique or an FET preamplifier stage may be used.

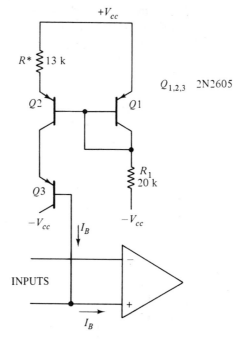

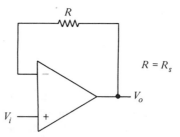

FIGURE 2.26   IMPROVED BIAS CURRENT CANCELLATION (NON-INVERTING MODE)

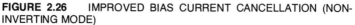

FIGURE 2.27   VOLTAGE FOLLOWER WITH BIAS CURRENT COMPENSATION FOR HIGH SOURCE IMPEDANCE

2.5   SUMMING AND DIFFERENCING AMPLIFIERS

In many analog signal processing applications, it is necessary to add two voltage signals. The circuit shown in Fig. 2.28 is useful in such cases. Since the voltage at the inverting input is essentially zero, a summing junction is formed: I_1

$= V_1/R_1$ and $I_2 = V_2/R_2$, also $I = I_1 + I_2$. The output voltage is the negative of IR:

$$V_o = -\left(\frac{R}{R_1} V_1 + \frac{R}{R_2} V_2\right)$$

Thus with $R_1 = R_2$, the output is a constant times the sum of the two input voltages. Note that the circuit allows for adding the two voltages in proportion (i.e. five times V_1 added to two times V_2) simply by scaling the resistors accordingly.

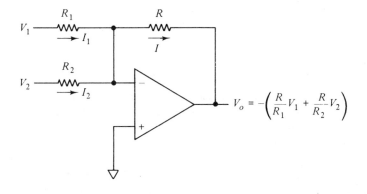

FIGURE 2.28   SUMMING AMPLIFIER (INVERTING)

Another very useful configuration, that of a difference amplifier, is shown in Fig. 2.29. Assuming the resistors to be matched, the output can be determined by considering the effect of each input separately and then adding the resulting outputs (this is the superposition theorem). Thus, with $V_2 = 0$, we have the inverting amplifier configuration: $V_o' = (-R_2/R_1)V_1$. With $V_1 = 0$, we have the non-inverting configuration with a voltage divider at the input: $V_o'' = (R_2/R_1)V_2$. The output is then:

$$V_o = V_o' + V_o'' = (R_2/R_1)(V_2 - V_1)$$

The difference between two inputs can then be amplified using the circuit in Fig. 2.29. It should be noted that a perfect match between the resistors in the inverting and non-inverting legs has been assumed. This is essential if the output is to be proportional only to the difference signal and not to the common-mode signal. To see the effect of resistor mismatch, let us consider 1% tolerance resistors used to give a gain of 100 ($R_1 = 1\,k, R_2 = 100\,k$). Further, consider the worst case when

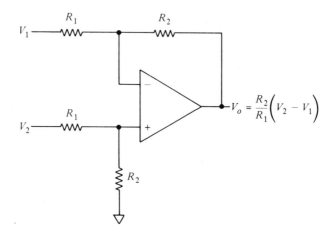

FIGURE 2.29   DIFFERENCE AMPLIFIER

each resistor is off by the maximum amount: the inverting gain limits are -98.02 and -102.02, obtained by letting one resistor be 1% high, the other 1% low. Similarly, the non-inverting gain limits are 97.03 and 101.03. We can then proceed to determine the differential and common-mode gains and the CMRR (see Sec. 2.2). For both sets of worst case gains (due to the resistors' mismatch), the resulting CMRR is about 40 dB. In such cases, the CMRR of the OP AMP itself is negligible. This example should illustrate the need for matched resistors.

2.6   INSTRUMENTATION AMPLIFIER

The difference amplifier has a low CMRR as well as providing a relatively low input impedance (the loading on V_1 is R_1). The improved differential amplifier shown in Fig. 2.30, also called an instrumentation amplifier, offers superior performance for the more critical applications like medical instrumentation.

Each input in Fig. 2.30 sees an extremely high impedance offered by the non-inverting inputs of OP AMPs 1 and 2. The gain of OA1 (assuming $V_2 = 0$) is the non-inverting gain $(R + R_1)/R$. Similarly, with $V_1 = 0$, the gain of OA2 is also $(R + R_1)/R$. Thus, the gains of OA1 and OA2 are matched if the two R_1's are matched; the two gains are adjustable with a single resistor, R. The third stage is a unity gain difference amplifier, again requiring matched resistors, R_2. With high accuracy digital multi-meters available readily and inexpensively, the matching process is relatively simple.

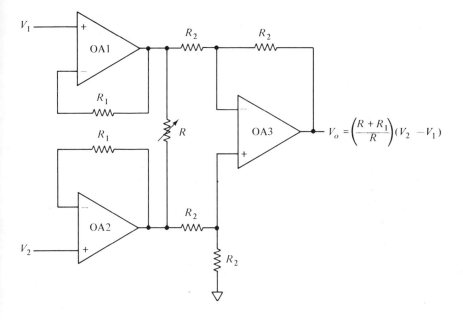

$$V_o = \left(\frac{R + R_1}{R}\right)(V_2 - V_1)$$

FIGURE 2.30   INSTRUMENTATION AMPLIFIER

This improved instrumentation amplifier is available in a single package, with all the resistors except R included. An example is the LH0036, a hybrid IC with thin film resistors trimmed to high accuracy, capable of operating from supply voltages as low as ± 1.5 V and providing 100 dB CMRR.

In medical applications, operation of the instrumentation amplifier from batteries, rather than from the ac power line, is extremely desirable to prevent possible shock to the patient.

2.7   CURRENT SOURCES AND SINKS

Another application for which OP AMPs can be used is in the conversion of a voltage signal (variable or fixed) into a current signal. Two of the many different versions are shown in Figs. 2.31 and 2.32.

Consider the precision current source first: for a negative input voltage, the output of the OP AMP swings in the positive direction, thus turning on the FET and bipolar output transistor, $Q1$ and $Q2$. With $Q2$ on, its collector current plus the FET drain current flow through R_1. Equilibrium is reached in the circuit when the drop across R_1 just equals V_i (the inverting and non-inverting inputs of the OP

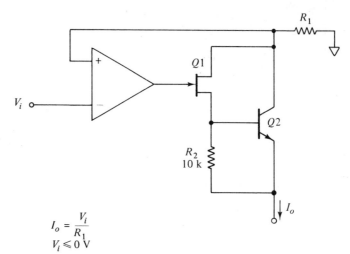

$$I_o = \frac{V_i}{R_1}$$
$$V_i \leqslant 0 \text{ V}$$

FIGURE 2.31   PRECISION CURRENT SOURCE

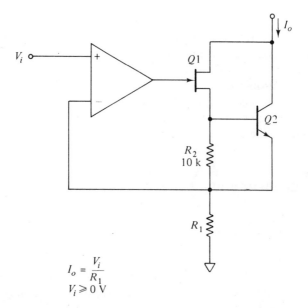

$$I_o = \frac{V_i}{R_1}$$
$$V_i \geqslant 0 \text{ V}$$

FIGURE 2.32   PRECISION CURRENT SINK

AMP at the same voltage). Then the current through R_1, and thus any load (referenced to the negative supply), must be V_i/R_1. The purpose of R_2 is to limit the FET current to about $I_{B2} + 0.06$ mA. By using the FET, which draws essentially zero gate current, the error in the conversion is much smaller than had a Darlington configuration been used.

Operation of the precision current sink, Fig. 2.32, is quite similar: for positive input voltages, the output again becomes positive turning on the FET and bipolar transistor, until equilibrium is reached with the voltage across R_1 equal to V_i. Under these conditions, the current through R_1, and also any load which is references to V^+, is V_i/R_1.

For both the source and sink circuit, the negative (or positive) supply must be high enough to accommodate the particular combination of input voltage and load. For example, with $R_1 = 1$ k the conversion is 1 mA/V; thus with a 10 V input and a 5 k load the supply should be approximately 50 V.

2.8 VOLTAGE REFERENCES

Most commonly, a zener diode is used to provide a fixed voltage reference for use in voltage regulator and other analog applications. However, the zener with the minimum temperature coefficient may not provide the desired reference voltage.

The positive and negative voltage references shown in Figs. 2.33 and 2.34 allow for continuous adjustment of the reference voltage, while at the same time providing current regulation for the zener. Referring to Fig. 2.33, the non-inverting input is at $V_oR_b/(R_a + R_b)$; this voltage also appears at the inverting input, which is clamped by the zener to the output. Thus:

$$V_o = V_Z + V_oR_b/(R_a + R_b)$$

or, solving for V_o:

$$V_o = V_Z \; \frac{R_a + R_b}{R_a}$$

The negative voltage reference output is similarly scaled by the voltage divider of R_a and R_b.

Once the desired output voltage is established by the appropriate choice of R_a and R_b, the zener current is set by R_1: $I_Z = (V_o - V_Z)/R_1$. This choice is made to minimize the temperature coefficient of the zener voltage.

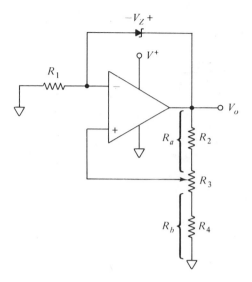

FIGURE 2.33   POSITIVE VOLTAGE REFERENCE

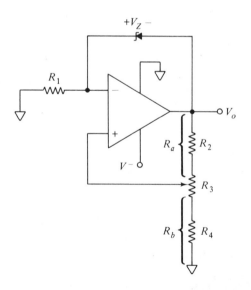

FIGURE 2.34   NEGATIVE VOLTAGE REFERENCE

2.9   PRECISION RECTIFIERS

The OP AMP can be used to idealize the conventional diode. Consider the circuit shown in Fig. 2.35. For a positive input, the OP AMP output is driven positive and causes the diode to conduct. Equilibrium is set up when the inverting input reaches the same voltage as the input, making output equal to the input. For negative inputs, the OP AMP output swings negative and saturates at a voltage roughly one volt lower than the negative supply. The diode is off, and the output voltage is zero. Thus the circuit provides rectification. Note that the minimum input voltage causing the diode to conduct is approximately 0.6 V divided by the open-loop gain—in effect, the OP AMP has reduced the diode cut-in voltage by a factor equal to the open-loop gain, thus idealizing the diode.

It should be obvious that the diode in Fig. 2.35 can be reversed, resulting in conduction for negative voltages.

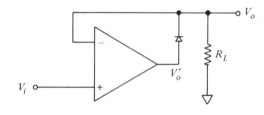

FIGURE 2.35   IDEAL DIODE

The precision rectifier circuit shown in Fig. 2.36 can provide the ideal rectification function and amplify the signal as well. For a negative input voltage, V_o' is driven positive and causes $D1$ to conduct. Thus negative feedback is established, and the voltage at the inverting input equalizes at zero. This is then the basic inverting amplifier configuration for negative inputs, with gain $-R_2/R_1$. (During this time $D2$ is reverse-biased and off.) For positive inputs, the OP AMP output swings negative, and $D2$ conducts. Equilibrium is reached when the current through R_1, $D2$ and into the OP AMP output is high enough to cause the inverting input to be at zero volts. $D1$ is now off, and there is no current through R_2; the output is then zero. The circuit provides inverting rectification with gain.

If the diodes in Fig. 2.36 are both reversed, the output is zero for negative inputs, and inverting gain is provided for positive inputs.

A precision full-wave rectifier (also called an absolute value) circuit is shown in Fig. 2.37. The circuit utilizes two OP AMPs, one connected in a

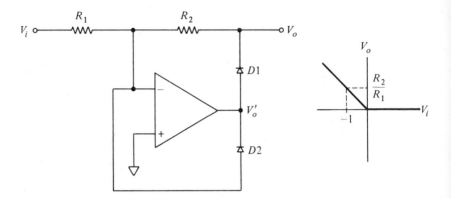

FIGURE 2.36   PRECISION RECTIFIER WITH GAIN

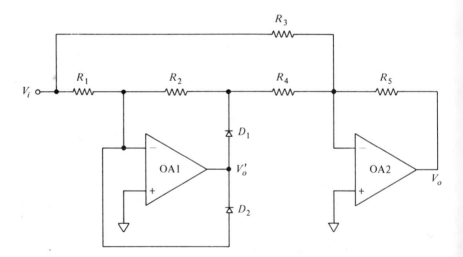

**FIGURE 2.37   PRECISION FULL-WAVE RECTIFIER (ABSOLUTE
VALUE) CIRCUIT**

summing mode, the other as a precision rectifier (just like the circuit of Fig. 2.36). The signal path to the output for V_i positive is only through R_3, since $D1$ is off, and the voltage applied to R_4 is zero. Thus for positive inputs:

$$V_o = -\frac{R_5}{R_3} V_i$$

When a negative input is applied there are two signal paths to the output, one through R_3 and the other through the precision rectifier formed by OA1. The input at R_4 is $-V_iR_2/R_1$. The total output is then:

$$V_o = -\left(\frac{R_5}{R_3} + \frac{R_5 R_2}{R_4 R_1}\right)V_i$$

If $R_4 = R_3$, then by adjusting the gain of the precision rectifier (R_2/R_1) to 2, the output for negative inputs is:

$$V_o = +\frac{R_5}{R_3} V_i$$

The output is always negative; thus the circuit performs full-wave rectification. Note that resistor ratios need to be set precisely to keep distortion to a minimum.

A non-inverting full-wave rectifier is constructed by reversing the two diodes in the circuit of Fig. 2.37, thus forcing a positive output no matter what the polarity of the input.

2.10   INTEGRATOR

If we note that in the inverting amplifier, the current through R_2 is independent of R_2 and set by V_i/R_1, we can readily see that a practical integrator can be constructed simply by replacing R_2 by a capacitor. This modification is indicated in Fig. 2.38. For constant current through the capacitor, the output voltage is the negative of the integral of the input voltage:

$$V_o = -\frac{1}{R_1C_o} \int_0^t V_i \, dt$$

A number of observations regarding the integrator circuit need to be noted: first, the circuit has essentially its dc gain equal to the open-loop gain of the OP AMP. Thus any dc component at the input (even as low as a few mV) will cause the output to ramp toward either the negative or positive supply voltage and eventually become saturated. The dc gain can be reduced by adding R_2 as shown (the dc gain is then $-R_2/R_1$), but linearity of the integrator is sacrificed. An alternate method is to ac couple the input waveshape, remembering to provide a dc path to the inverting input to allow for bias current.

The obvious limitations on the accuracy are the quality of the capacitor (it should have an extremely low leakage resistance) and the bias current of the OP

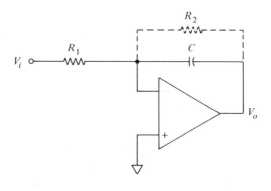

FIGURE 2.38   INTEGRATOR

AMP. Thus for integrator applications, OP AMPs with very low bias currents should be used. In any case, the integrating current should be at least two orders of magnitude larger than the input bias current to minimize any errors at the output.

2.11   SCHMITT TRIGGER

The basic Schmitt trigger circuit utilizes two transistors with positive or regenerative feedback. The design of this circuit is by no means a simple task. However, the basic idea of that circuit can be implemented with an OP AMP. One of the possible configurations is shown in Fig. 2.39. For the moment, consider V_R to be zero. Positive feedback is applied by taking a fraction of V_o, accomplished by the voltage divider formed by R_1 and R_2, and returning it to the non-inverting input. Since there is no negative feedback, the output voltage will be at either its positive or negative saturation value (near V_{cc}). Let us assume $V_o = +V_s$. The positive feedback voltage at the non-inverting input is then:

$$V_x = \frac{R_2}{R_1 + R_2}\, V_s = \beta\, V_s$$

where β is defined as the voltage divider ratio (not to be confused with the transistor β). This set of conditions can only exist if the input voltage is less positive than V_x. Now if the input voltage is allowed to equal V_x and exceed it just slightly, the effective input to the OP AMP, $(V_1 - V_2)$, becomes positive and is multiplied by the large negative number that is the open-loop gain. This drives

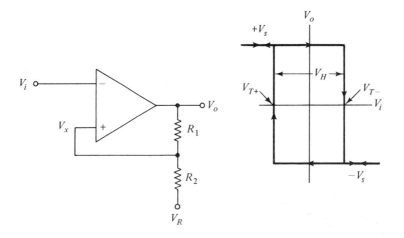

FIGURE 2.39   INVERTING SCHMITT TRIGGER

the output negative, reducing V_x, which increases the effective input voltage, and drives the output even more negative. Equilibrium is reached when $V_o = -V_s$ with V_i more positive than βV_s. Thus we see that the negative threshhold voltage at the input to drive the output negative is:

$$V_{T-} = \beta V_s$$

Once the output is negative, $V_x = -\beta V_s$, returning the input below βV_s does *not* cause the output to switch back to V_s. In fact the input must be made slightly more negative than $-\beta V_s$ before the effective OP AMP input becomes negative, thus driving the output positive. The positive threshold voltage is then:

$$V_{T+} = -\beta V_s$$

Again, once the output is positive, the input must be brought more positive than βV_s before the output switches back to the negative saturation value.

If we consider the addition of a reference voltage as shown, the two threshold voltages are modified. By calculating the effects of V_o and V_R with each acting separately, then adding the results (i.e. using the superposition theorem), we can readily see that:

$$V_{T-} = \beta V_s + (1 - \beta) V_R$$

and

$$V_{T+} = -\beta V_s + (1 - \beta) V_R$$

The circuit exhibits hysteresis—that is, the output has two possible states for one input voltage. Thus, predicting the output for a given input voltage may require knowing the previous state of the input.

Note that in many applications it is necessary to have the two threshold voltages set precisely. The threshold voltages, however, are a function of the OP AMP output saturation voltage, which at best is rather unpredictable, varying from one OP AMP to another (even of the same type). To achieve somewhat better and more predictable performance, the circuit is modified by adding a series dropping resistor R and two back-to-back zeners, as shown in Fig. 2.40. Basic operation is unchanged, except that the output voltage is either at + or − $V_Z + V_D$. The zeners are always chosen to have their zener voltage at least a few volts lower than the supply voltage.   The resistor R limits the zener current to $(V_s - V_Z - V_D)/R$, and is chosen accordingly.

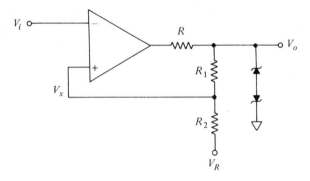

FIGURE 2.40   SCHMITT TRIGGER WITH PREDICTABLE PERFOR-MANCE

The improvement offered by this circuit is that now the threshold voltages are:

$$V_{T-} = \beta \, (V_Z + V_D) + (1 - \beta) \, V_R$$

and

$$V_{T+} = - \, \beta \, (V_Z + V_D) + (1 - \beta) \, V_R$$

Note that both threshold voltages are now independent of the OP AMP output saturation voltage and are set by the zener and reference voltages, providing more reliable operation.

The reference voltage should be derived from a zener or other low source impedance network; for example see the voltage reference circuits of Figs. 2.33

and 2.34. For low-valued reference voltages, a potentiometer can be used to step the zener voltage down to the desired value. The wiper of the potentiometer is then buffered with an OP AMP voltage-follower circuit.

A non-inverting Schmitt trigger is shown in Fig. 2.41. Note that the input voltage and the reference have been interchanged. The major difference between the non-inverting and inverting configurations is the high input impedance of the inverting circuit, while in the non-inverting case V_i is loaded by R_1 and R_2. The back-to-back zener circuit can be added to the non-inverting configuration, with a corresponding improvement in predictability of the threshhold levels:

$$V_{T+} = \frac{V_R + \beta V_s}{1 - \beta}$$

$$V_{T-} = \frac{V_R - \beta V_s}{1 - \beta}$$

where, for the circuit with the back-to-back zeners, V_s is replaced by $V_Z + V_D$ as before.

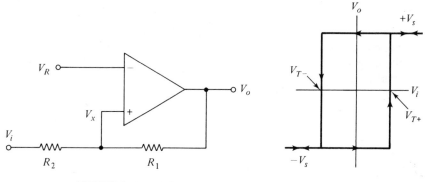

FIGURE 2.41   NON-INVERTING SCHMITT TRIGGER

The Schmitt trigger circuits are extremely useful in many applications: converting analog inputs to square waves, in temperature controllers, and as part of waveshape-generating circuits. A brief example will illustrate the application in thermal control. A temperature sensor provides an input of 2 V at 20°C. In order to maintain the temperature between 19° and 21°C (input between 1.95 and 2.05 V, respectively), the heater should be turned on (negative voltage required) when the temperature reaches 19°, and turned off (positive voltage required) when the temperature reaches 21°. This translates to $V_{T-} = 2.05$ V and $V_{T+} = 1.95$ V.

Note that the hysteresis voltage is $V_H = V_{T-} - V_{T+} = 100$ mV. Using the inverting Schmitt trigger with $V_Z + V_D = 10$ V:

$$V_H = 100 \text{ mV} = 2\beta(10 \text{ V})$$

Thus: $\beta = 0.005$. We might choose $R_1 = 10$ k, and trim R_2 to 50 Ω by using a 100 Ω potentiometer.

The sum of the two threshold voltages is 4 V, and from the equations we can determine V_R:

$$V_R = \frac{4 \text{ V}}{2(1 - \beta)} \cong 2.01 \text{ V}$$

To provide this voltage, we might use a higher voltage zener (6.8 V, for example) with a 10-turn 10 k potentiometer for fine adjustment, followed by an OP AMP voltage follower to provide the low source impedance required for V_R.

2.12   SQUARE-WAVE AND PULSE GENERATORS

By adding a resistor and a capacitor to the inverting Schmitt trigger, as shown in Fig. 2.42, a simple square-wave oscillator is constructed. Note that the Scmitt trigger has no reference voltage, and thus has symmetrical operation centered around ground: $V_{T-} = +\beta V_s$ and $V_{T+} = -\beta V_s$.

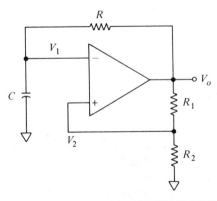

FIGURE 2.42   SIMPLE SQUARE-WAVE GENERATOR

To determine the timing, note that when the output is $+V_s$, the capacitor charges toward this value. However, when the capacitor voltage reaches V_{T-}, the output switches to $-V_s$. Now the capacitor charges toward $-V_s$ until the capacitor voltage reaches V_{T+}, and the output once again switches to $+V_s$. The capacitor voltage alternates between V_{T-} and V_{T+}. Recall that in general, a capacitor charging through a resistor will reach a voltage V_x at time t_x if its initial (at $t = 0$) voltage is V_i, and the voltage it is charging to is V_f. In equation form, this is:

$$V_x = V_f - (V_f - V_i)\exp(-t_x/RC)$$

For the square-wave generator, we have $V_x = V_{T-}$ when $t = T/2$ (where T is the period), and $V_f = V_s$, while $V_i = -\beta V_s$. Thus substituting in the above equation and solving for T, we get:

$ln = Log n$

$$\boxed{T = 2RC \; ln\left(1 + \frac{2R_2}{R_1}\right)}$$

This result is valid only if the OP AMP output saturation voltage is the same in both the positive and negative directions. Then, as can be seen above, the period is essentially independent of the output voltage swing. To make certain that the positive and negative output voltages are numerically equal, the back-to-back zener diode circuit, as shown in Fig. 2.40, can be included in the square-wave generator circuit, with R tied to the top of the zeners instead of the OP AMP output. (For identical zeners, the period remains as given above.)

The circuit of Fig. 2.42 gives a duty cycle of 50%, i.e. the portion of the total period with the output positive is the same as that with the output negative. By providing different time constants for charging and discharging the capacitor, the duty cycle can be adjusted to values other than 50%. This is accomplished by inserting diodes in series with the timing resistors, as shown in Fig. 2.43. With the output positive, the capacitor charges through R_a and $D1$; while for negative outputs, the charging path is through R_b and and $D2$. The foregoing discussion of timing determination still applies, only the time constants are different. Thus, assuming the diode resistance negligible with respect to either R_a or R_b, we can show that:

$$T = (R_a + R_b)C \; ln\left(1 + \frac{2R_2}{R_1}\right)$$

If T_1 is the time that the output is positive, then:

$$T_1 = R_aC \; ln\left(1 + \frac{2R_2}{R_1}\right)$$

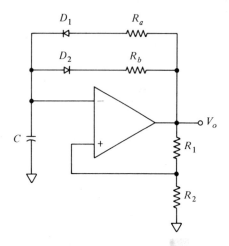

FIGURE 2.43   SQUARE-WAVE GENERATOR FOR OTHER THAN 0.5
DUTY CYCLE

Similarly, if T_2 is the time for a negative output, then:

$$T_2 = R_b C \; ln \left(1 + \frac{2R_2}{R_1} \right)$$

The duty cycle is then the ratio of T_1 to T:

$$Duty \; cycle = \frac{T_1}{T_1 + T_2} = \frac{R_a}{R_a + R_b}$$

To express the duty cycle in percent, multiply the ratio above by 100.

The capacitor voltage as well as the output voltage for the square-wave generator are shown in Fig. 2.44. This timing diagram is for the case where R_b is greater than R_a. Note that when $R_a = R_b$, $T_1 = T_2 = T/2$ and operation is essentially the same as for the basic square-wave generator.

If the fixed resistors of Fig. 2.43 are replaced by a potentiometer, as shown in Fig. 2.45, a pulse generator with continuously variable duty cycle results. (Although all pulse generators provide square-wave outputs, generally if the duty cycle is other than 50%, the square-wave generator is termed a pulse generator.)

The square-wave generators of Figs. 2.42, 2.43, and 2.45 have a basic limitation in the maximum frequency of operation. The limitation is one due to the slew-rate of the OP AMP. To obtain a square wave with a rise or fall time which is not a significant proportion of the period, the circuits must be operated

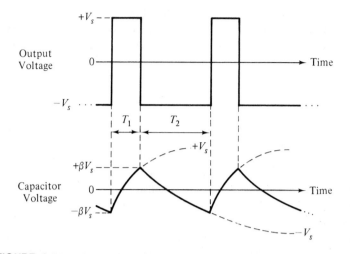

FIGURE 2.44 WAVESHAPES FOR SQUARE-WAVE GENERATOR
$(R_a < R_b)$

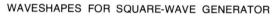
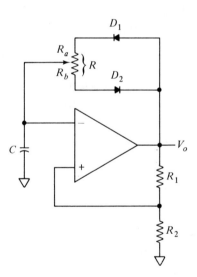

FIGURE 2.45 SQUARE-WAVE GENERATOR WITH ADJUSTABLE
DUTY CYCLE

at repetition rates that are relatively low. For example, if a 741 OP AMP is used
with a slew rate of 0.5 V/μs, and the output voltage swing is maintained between
+ 10 V and − 10V, the rise (or fall) time is approximately 20V/SR = 40 μs.

Thus for a good square wave, with the sum of the rise and fall time less than 5% of the period, the minimum period is $20 \times 2 \times 40 \ \mu s = 1.6$ ms. This gives a maximum frequency of 625 Hz. It should be obvious that this frequency limitation is minimized if the output voltage swing is reduced, or if using OP AMPs with a high slew-rate.

2.13   ACTIVE FILTERS

One of the most important applications of OP AMPs is in the realm of active filters. By definition, an active filter is one that utilizes an active device (transistor, FET, OP AMP) in addition to R, L, and C to provide the desired frequency response. The use of active filters for low-frequency applications is dictated by the fact that inordinately large values of L and/or C are required. OP AMPs are used to increase the effective values of L and C.

An exhaustive discussion of active filters will not be attempted. Many good references exist, treating only this area. In this section we shall examine the basic techniques used in applying OP AMPs to filters.

Consider first the synthetic inductor circuit of Fig. 2.46. It utilizes an OP AMP in the voltage follower configuration. Devices, such as the 110 (310) voltage follower are available with the connection between the inverting input and the output provided internally. The advantage of using a voltage follower like the 110 in such applications is that it is designed to have a wider frequency response, higher slew-rate, and lower bias current than a typical OP AMP connected externally as a voltage follower.

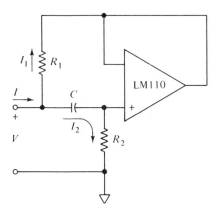

FIGURE 2.46   SYNTHESIZED INDUCTOR ($L = R_1R_2C$)

The OP AMP in Fig. 2.46 forces the same voltage on the top of R_1 as exists across C. Thus $I_1R_1 = I_2/sC$. Also $I=I_1+I_2$. To find the impedance at the input, we evaluate the ratio of V to I. The applied voltage is:

$$V = I_2(1/sC + R_2)$$

Using the relationships among I, I_1, and I_2, we see that:

$$Z = \frac{V}{I} = \frac{R_1 + sR_1R_2C}{1 + sR_1C}$$

For frequencies significantly below $1/2\,\pi R_1 C$, the term sR_1C is negligible with respect to 1, so:

$$Z \cong R_1 + sR_1R_2C = R_s + sL$$

where $R_s = R_1$ and is the effective series resistance of the synthesized coil, and $L = R_1R_2C$ is the effective inductance of the synthesized coil. For example, $R = 100\Omega$, $R = 10M$, and $C = 0.1\ \mu F$ gives a coil with a series resistance of 100Ω and an inductance of 1000 H. The maximum usable frequency is $1/40\pi R_1 C$ or approximately 80 Hz (for less than 5% error in neglecting the 1 in the discussion above). Note that for the numbers used, the Q of the coil at 10 Hz is $628(\Omega L/R_s)$. This gives significantly better results than can be obtained from actual inductors, not to mention the advantage of avoiding the gigantic physical size of an actual 1000 H coil.

The capacitance multiplier circuit shown in Fig. 2.47 uses a voltage follower in much the same manner. The capacitor voltage is forced on the top of R_1 by the voltage follower. Thus, the voltages across R_1 and R_2 must be equal: $I_1R_1 = I_2R_2$. Also, $I = I_1+I_2$ and $V = I_2(R_2+ 1/sC)$. After appropriate substitutions, and restricting operation to frequencies significantly lower than $1/2\,\pi R_2 C$:

$$Z = \frac{V}{I} = \frac{1}{sC'} + \frac{R_1R_2}{R_1 + R_2}$$

where $C' = C(1 + R_2/R_1)$. If R_2 is large compared to R_1, the effective impedance offered by the circuit in Fig. 2.47 is that of a capacitance C' and effective series resistance of approximately R_1. For example: $R_1 = 1\ k$, $R_2 = 10\ M$, and $C = 1\ \mu F$ gives an effective capacitance of 10,000 μF and an effective series resistance of 1 k. For less than 5% error, operation is limited to frequencies at or below $1/40\pi R_2 C$ or approximately 8 Hz for this example.

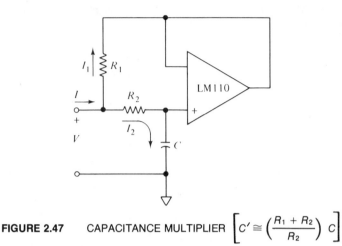

FIGURE 2.47   CAPACITANCE MULTIPLIER $\left[c' \cong \left(\dfrac{R_1 + R_2}{R_2} \right) c \right]$

Other filter applications involve the use of more than a single OP AMP. For example, the so-called state variable filter, which provides low-pass, band-pass, and high-pass outputs simultaneously, requires the use of three OP AMPs. In such cases the use of triple OP AMP ICs (the L144 being an example) is quite convenient. In addition, some IC manufacturers offer universal active filter building block ICs (the AF100, for example), which in addition to OP AMPs, include resistors and capacitors to enable the realization of complex filtering functions with a minimum of external components.

2.14   FREQUENCY COMPENSATION AND SLEW-RATE

We have already mentioned the OP AMP open-loop frequency response and its slew-rate. Now we want to examine the effects of these two quantities in some detail.

In just about all the applications discussed in previous sections, the OP AMP is used with differing amounts of negative feedback. The effect of this feedback on the closed-loop frequency response is of major consequence.

First consider the frequency response of internally compensated OP AMPs. The 741 is typical (see Fig. 2.9). We see that at 1 MHz where the open-loop gain is 0 dB (unity), the phase shift is approximately $-100°$. This gives a phase margin of 80°, and assures stable operation (no oscillations) under all negative feedback conditions. [For a detailed discussion of feedback amplifiers and gain and phase margins, see Michael Cirovic, *Semiconductors: Physics, Devices and Circuits* (Englewood Cliffs, N.J.: Prentice-Hall, Inc., 1971), Chap. 8, pp. 295–329.]

However, this phase margin may be significantly smaller, depending on the internal device parameters which do vary from one OP AMP to another. Remember that the plots are only typical. With purely resistive (negative) feedback, no stray capacitance anywhere, and a purely resistive load, the OP AMP is unconditionally stable. In practice, stray capacitance, especially at the inverting input, may cause the amplifier to oscillate. In addition, the phase shift caused by certain load conditions (heavily capacitive loads) may also cause oscillations. As if this were not enough, inductive effects in the power supply may themselves cause the amplifier to break into oscillation. So we see that even an OP AMP that is internally compensated may need external compensation under certain conditions.

In most cases the problems outlined above are easily overcome. As a general rule, it is a good idea to operate OP AMPs with no more bandwidth than necessary. For example, if the inverting amplifier (shown in Fig. 2.18) is used to amplify a dc signal or one of very low frequency, it should be compensated for an upper 3 dB frequency just sufficient to handle the requirements. For a gain of 100 (40 dB), and with a gain-bandwidth product of 1 MHz, the amplifier bandwidth without compensation is 10 kHz. If the largest frequency of the input signal is 100 Hz, there is obviously excessive bandwidth. The closed loop bandwidth can be easily adjusted downward, simply by adding a capacitor in parallel with R_2. The new upper 3 dB frequency is then:

$$f_{3\,dB} = \frac{1}{2\pi R_2 C}$$

Thus, if in our example $R_2 = 100$ k, then a 15 pF capacitor will give a bandwidth of approximately 100 Hz. The reduction in bandwidth compensates for any stray capacitance at the inverting input.

The second general rule is to bypass the power supply terminal as close to the V_{cc} pins of the OP AMP as possible. Typically, 0.01 μF or larger capacitance would be sufficient to eliminate the possibility of oscillation due to power supply inductance. Bypassing of the power supply is especially important in applications where a number of OP AMPs are run from the same supplies.

For heavy capacitive loads, the simplest remedy is to isolate the load by inserting a small resistance between the point common to the load and the actual OP AMP output.

The internally compensated OP AMP is a practical choice in all dc and very low frequency applications. However, the uncompensated OP AMP offers greater flexibility where wider bandwidth and/or higher slew-rate are needed.

Two basic frequency compensation schemes are used for the uncompensated OP AMP. They are the dominant-pole scheme and the feed-forward compensation. Others are possible, but the two mentioned are sufficient. The dominant-pole compensation scheme is the one used by manufacturers for the internally compensated OP AMPs and involves placing a low value capacitor, typically 30 pF, between the input and output of a voltage gain stage. This provides for a frequency response very similar to that of the 741, a rolloff of 20 dB per decade (6 dB/octave) and a gain-bandwidth product of approximately 1 MHz. A typical circuit and its frequency response characteristics are shown in Figs. 2.48 and 2.49 respectively. Consider the following example: a 301 is used in the inverting configuration with a closed-loop gain of -100. Using a 30 pF compensating capacitor gives us a bandwidth of approximately 10 kHz. But note that the loop gain is 60 dB and has a gain-bandwidth product of approximately 10 kHz, providing for a phase margin of 90°. The same amplifier when operated with a 3 pF compensating capacitor has a closed-loop bandwidth of approximately 100 kHz; loop-gain is the same and so is the phase margin. Thus, the bandwidth has

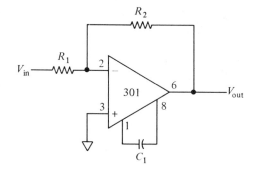

FIGURE 2.48   STANDARD FREQUENCY COMPENSATION

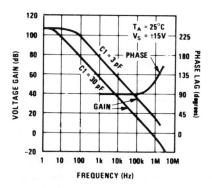

FIGURE 2.49   OPEN-LOOP FREQUENCY RESPONSE FOR STANDARD COMPENSATION

been increased by a factor of 10 (the slew-rate also), while stability has not been jeopardized. A word of caution: with the 3 pF capacitor, the resulting amplifier would probably oscillate should the closed-loop gain be decreased to near 0 dB.

The bandwidth can also be reduced for applications where it is not required. For example, if a 300 pF compensating capacitor is used, the gain-bandwidth product would be approximately 100 kHz, with a resulting reduction of closed-loop bandwith.

A further improvement in bandwidth can be attained with feed-forward compensation. We note that the active loads for the differential amplifier (internal to the OP AMP) are *PNP* transistors. These provide excess phase shift at higher frequencies. This is also accompanied by a reduction in open-loop gain. To overcome this problem, a capacitor is connected between the inverting input and a point beyond the *PNP* transistors. In this manner, at high frequencies where the *PNP* transistors degrade performance, they are bypassed and the input signal is fed forward ahead of the *PNP*s. The circuit is shown in Fig. 2.50. Note that for

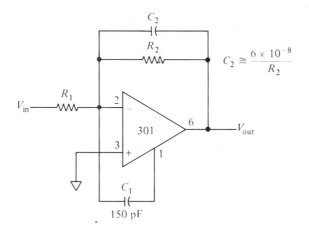

FIGURE 2.50   FEED-FORWARD FREQUENCY COMPENSATION

stability, an additional capacitor C_2 is placed in parallel with R_2. This capacitor is chosen to give a break frequency (together with R_2) at 2 or 3 MHz:

$$C_2 = 60/R_2 \qquad (C_2 \text{ in pF, } R_2 \text{ in k})$$

The resulting open-loop frequency response is shown in Fig. 2.51. The rolloff is at approximately 20 dB/decade for frequencies below 1 MHz. Note that for a closed-loop gain of 40 dB, the bandwidth is in excess of 200 kHz, while the phase margin is approximately 80°.

A direct comparison of the open-loop frequency response for the standard and feed-forward compensation schemes, shown in Fig. 2.52, shows an improvement factor of about 10 in bandwidth as well as slew-rate (not shown) for the feed-forward scheme over the standard compensation technique. This bandwidth improvement is available for all closed-loop gain values.

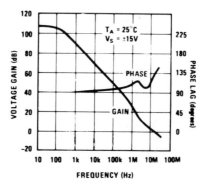

FIGURE 2.51   OPEN-LOOP FREQUENCY RESPONSE WITH FEED-FORWARD COMPENSATION

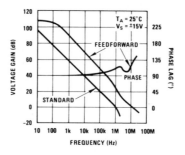

FIGURE 2.52   COMPARISON OF THE OPEN-LOOP FREQUENCY RE-SPONSES FOR STANDARD AND FEED-FORWARD COMPENSATION

Thus the type as well as the amount of compensation determine the small-signal bandwidth for a given gain. Once the amplifier has been compensated, the slew-rate is set. The large signal behavior is primarily determined by the slew-rate rather than by the bandwidth. For example, let us assume a closed-loop gain and bandwidth of 40 dB and 10 kHz respectively, and a slew-rate of 0.5 V/μs. Further, assume an sinusoidal input signal of 100 mV peak. With the OP AMP operating with ±15 V supplies, we would expect a 10 V peak output over the specified bandwidth of 10 kHz, with the amplitude decreasing at frequencies of 10 kHz and above, but a sinusoidal signal nevertheless. However such would not be the case if we carried out the experiment at 10 kHz; we would observe a triangular waveshape at the output. The reason for this distortion of the signal is that the amplifier is slew-rate limited.

Consider the output waveshape to be given by:

$$V_o = V_{peak} \sin wt$$

Then the rate of change of the output voltage is:

$$\frac{dV_o}{dt} = \Omega \, V_{peak} \cos wt$$

This time rate of change of the output is a maximum when $t = 0$, and the OP AMP maximum time rate of change at the output is the slew-rate (SR). Equating the two, we can determine the maximum frequency of a sinusoidal signal that the amplifier will pass without distortion:

$$f_{max} = \frac{SR}{2 \pi V_{peak}}$$

If we use the numbers from the example above, we see that the maximum frequency due to slew-rate limitations is 7.9 kHz, although the bandwidth is 10 kHz.

In general, we see that the maximum sinusoidal signal frequency that can be handled without distortion is only a function of the OP AMP slew-rate and the peak amplitude of the output signal.

This same slew-rate limitation plagues most circuits discussed previously. We have already discussed it in the case of the Schmitt trigger and square-wave generators. Slewing at higher frequencies is also a serious problem for the ideal diode circuit of Fig. 2.35. This might not be expected at first glance since the circuit is used to rectify extremely small voltages. However, when the input to the ideal diode circuit shown in Fig. 2.35 is negative, note that the output of the OP AMP V_o' swings to the negative saturation voltage (approximately $-V_{cc}$). Thus when the input is made positive, the OP AMP output is required to also swing positive—a large change in voltage which occurs at the slewing rate. For any but the lowest frequencies, the precision rectifier circuit is superior, especially if an uncompensated OP AMP with feed-forward compensation is used.

2.15   NON-IDEAL OPERATION

In discussing the applications of OP AMPs, we have by and large assumed the OP AMP to be ideal. We have already seen the effects of non-ideal bandwidth and slew-rate. We now consider the effects of non-infinite open-loop gain, input resistance, and non-zero output resistance.

Consider first the inverting amplifier configuration shown in Fig. 2.18. If the open-loop gain is finite, then there is a voltage between the OP AMP input terminals equal to V_o/A_{OL}. Also, the current $I = (V_i - V_o/A_{OL})/R_1$. Thus, solving for the closed-loop gain, we obtain:

$$\frac{V_o}{V_i} = -\frac{R_2}{R_1}\left[\frac{1}{1 - \frac{1}{A_{OL}}\left(\frac{R_1 + R_2}{R_1}\right)}\right]$$

An example will illustrate the inherent error in assuming ideal operation. Assume $R_1 = 1\ \text{k}, R_2 = 100\ \text{k}$ and $A_{OL} = -50,000$ (not at all untypical). The approximate gain is -100; the exact gain is -99.8, giving an error of 0.2%. The uncertainty in the resistors is usually larger—thus it is a good assumption to say that the gain is infinite.

Similar analysis yields the exact gain for the non-inverting configuration:

$$\frac{V_o}{V_i} = \frac{R_1 + R_2}{R_1}\left[\frac{1}{1 - \frac{1}{A_{OL}}\left(\frac{R_1 + R_2}{R_1}\right)}\right]$$

Note that the correction factor is the same for the inverting and non-inverting cases, yielding similar results and conclusions.

Consider next including the effect of the OP AMP input resistance R_i. Then the current through R_2 is not equal to the current through R_1; the difference is $V_o/A_{OL}R_i$. Using this relationship and simplifying, we obtain:

$$\frac{V_o}{V_i} = -\frac{R_2}{R_1}\left[\frac{1}{1 - \frac{1}{A_{OL}}\frac{R_1}{R_2} + \left(\frac{R_i + R_2}{R_i}\right)}\right]$$

For typical values $A_{OL} = -50,000$ and $R_i = 2\ \text{M}$, the gain error is less than 0.2%.

Lastly, we consider the effect of a non-zero output impedance R_o. With R_o included, the output voltage is reduced by IR_o. Using this fact and making the appropriate substitutions, we obtain:

$$\frac{V_o}{V_i} = -\frac{R_2}{R_1}\left[1 - \frac{R_o/R_2}{1 - \left(\frac{A_{OL}R_1}{R_1 + R_2}\right)}\right]$$

Typically, $R_o = 75\Omega$, and using the same value for the open-loop gain and gain setting resistors, the gain error is extremely small, less than 0.01%.

In most applications the effects of the open-loop gain and the input and output resistance may safely be neglected.

2.16   SPECIAL OP AMPs

In preceding sections we have already mentioned some special OP AMPs: the programmable types, high slew-rate types, voltage-follower types, and low bias current types. Here we want to add to the list: FET input, MOSFET, and single-supply types which meet very specific and special needs.

In many applications, a very high input impedance as well as an extremely low input bias current are required. In such cases two options exist: we can elect to build an FET input preamplifier for a conventional bipolar OP AMP like the 741, or we can use one of the commercially available FET input OP AMPs. To build an FET input preamplifier, a matched dual FET like the Siliconix U401 is used, as shown in Fig. 2.53. The CR043 current regulator (actually an FET) is used as the current source for the FET differential amplifier; offset adjustment is provided by the 1 k potentiometer. (More details on this circuit are available in AN74-3 from Siliconix.)

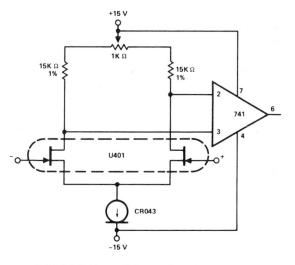

FIGURE 2.53   FET INPUT PREAMPLIFIER

OP AMPs like the National LF156 provide the FET input in a single package. In addition, RCA type CA3130 OP AMP incorporates MOSFETs at both the input and output, while the CA3140 has a MOSFET input and bipolar output (called a BIMOS OP AMP). The advantage of the MOSFET inputs, in addition to the very high input impedance and low input bias current (1.5 TΩ and 10 pA typical), is the increase in the common-mode input-voltage range. The input MOSFETs remain biased even when the input is brought to either supply voltage. The CA3130 output can also swing to within a few mV of either supply voltage, whereas bipolar transistor output OP AMPs are limited to within a few diode drops of V_{cc}. The CA3140 is a pin-for-pin replacement for a 741, with obvious improvements in input impedance and bias current as well as slew-rate.

Another special application for OP AMPs is that requiring only one supply. The LM324 is an example of such an OP AMP. (The LM324 contains four identical OP AMPs.) The circuit is shown in Fig. 2.54. Unlike the FET OP AMPs, the 324 is a low-cost general-purpose amplifier intended for single supply applications. It can operate on any supply voltage between 3 and 30 V. Note that the use of *PNP* differential input transistors results in a bias current out of the input. Also, the use of *PNP* transistors at the input allows for a common-mode input anywhere from ground to approximately three diode drops lower than V_{cc}. In addition, the output *PNP* allows the output voltage to reach within typically 100 mV of ground. In using the 324 in single supply applications, one of the main things to remember is that the output cannot go negative, so that a dc inverting amplifier is not possible. For ac inverting amplifiers, the non-inverting input should be biased at $V^+/2$ and the signal capacitively coupled. Note that the 324 can be operated from a 5 V supply, making it ideal for use in the analog section of basically TTL systems (see Chapter 4).

To summarize, we have examined a few representative OP AMPs in some detail, as well as a limited number of OP AMP applications. One thing should be apparent: the monolithic OP AMP is an extremely versatile IC—a fact enhanced by the ready availability of many different types of OP AMPs to fulfill almost any need.

REVIEW QUESTIONS

2.1   What is the typical open-loop gain for monolithic OP AMPs?

2.2   What is the difference between uncompensated and internally compensated OP AMPs? Give an example of each type.

2.3   How do programmable OP AMPs like the LM4250 differ from general purpose OP AMPs like the 741? Enumerate the programmable parameters.

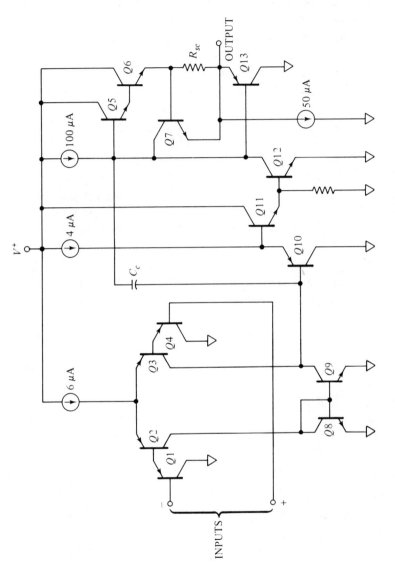

FIGURE 2.54   SCHEMATIC FOR LM324 (ONE OF FOUR IDENTICAL)

71

2.4   A 741 is used in the circuit of Fig. 2.18. The desired midband gain is -10.

   a. Specify the ratio of resistors R_1 and R_2, as well as an allowable range for
      the resistor values.

   b. What is the closed-loop upper 3-dB frequency?

   c. When the input is grounded, what is the largest output that can be
      expected due to the input offset voltage?

   d. If the resistor ratio in a. above is exact, what is the gain error due to the
      non-infinite open-loop gain of the 741?

   e. If the input voltage is 1 V dc, the output will be -10 V. Under these
      conditions, estimate the actual voltage across the input of the OP AMP.
      (*Note*: Ideally, this voltage should be zero, but due to the non-infinite
      open-loop gain, it will not be zero.)

   f. If the input is sinusoidal with a 1 V peak value, determine the highest
      frequency which can be amplified without distortion. (*Hint*: Determine
      the frequency at which the output is slew-rate limited.)

2.5   Make a comparison of the inverting and non-inverting amplifier configura-
       tions, i.e. list similarities and differences in their characteristics.

2.6   An inverting amplifier with a gain of 40 dB is required. By consulting data
       sheets (Appendix), compare the available bandwidth using an internally
       compensated 741, with an LM301 which has (a) single (dominant) pole
       compensation, (b) lead-lag compensation, (c) feed-forward compensation.

2.7   In a micropower application, an LM4250 is used with ± 1.5 V supplies.
       The circuit is to function between $0°$ and $60°C$ with I_{set} of 10 μA. List
       the limits on all pertinent amplifier characteristics.

2.8   A 741 is used in the inverting configuration shown in Fig. 2.18, with $R_1 = 1$
       MΩ and $R_2 = 2$ MΩ. Neglecting the effects of offset voltage, what would
       the output be (due to input bias current) when the input is zero? (*Note*:
       Ideally, the output should be zero. Use a typical open-loop gain of 90 dB
       and a maximum bias current of 0.5 μA.)

2.9   If Question 2.8 is modified to include the bias current compensating resistor
       R as in Fig. 2.22, what value should R have? Assuming the maximum input
       offset current, and neglecting the effects of offset voltage, what is the
       worst-case output voltage when the input is zero?

2.10   The zener in Fig. 2.34 is rated at 6.8 V at 10 mA. The desired output
       reference voltage is -10 V. A 741 is used. Specify all resistor values.

2.11   The precision rectifier of Fig. 2.36 is to be implemented with a gain of -10
       using a 741. The sine-wave input will have a peak value of 1 V or less.
       Specify circuit values and the highest frequency at the input if the output is

to have negligible distortion. (Use 0.5 V as the diode on voltage and a slew-rate of 0.5 V/μs. Assume the output distortion to be negligible if the OP AMP is slewing no more than 5% of the time.)

2.12 Implement a circuit to give a high output when the input exceeds 2 V and a low output when the input falls below 0.8 V. Specify all circuit values. (*Hint:* Use the circuit in Fig. 2.40 with a voltage follower to provide the reference; additional OP AMPs may be necessary to provide proper output phasing.)

2.13 The circuit in Fig. 2.42 is to provide a 100-Hz square wave, using a 0.22 F capacitor. Specify resistor values. If a 741 is used (output swings between +12 and −12 V), specify the maximum rise and fall times of the output.

2.14 Compare the performance (bandwidth) of a 301 OP AMP with single-pole and feedforward compensation with closed-loop gains of: (a) 60 dB, (b) 40 dB, and (c) 20 dB.

2.15 Give an example of when it would be advantageous to use a 324 OP AMP instead of a 741. (Disregard the fact that the 324 IC contains four OP AMPs, while the 741 is a single OP AMP.)

3

Voltage
Regulators

Whether we consider discrete-component or IC circuits, all circuits utilizing active devices require a dc power supply to provide bias. In discrete-component designs, and to a lesser degree in IC designs, the power supply needs to be well regulated since any variation in the operating point of any device results in a change in all parameters. Thus the IC voltage regulator plays an important role in any electronic system, large or small. It offers ease of use, excellent performance, and low cost.

3.1   BASIC SERIES REGULATOR

In the simplest terms, a voltage regulator is a functional block which accepts a varying dc voltage at the input and provides an essentially constant output dc voltage over a wide range of load conditions. There are two regulator types: series and shunt regulators. Of the two, the series regulator offers superior performance and is by far the more common. The series regulator, in one version or another, is used in just about all applications except for extremely high power supplies. In such cases, a switching regulator is used due to its very high power efficiency. Although the series regulator is relatively inefficient, since it is used in low power applications, the resulting power loss is far outweighed by the simplicity, lower cost, and better performance.

The simplest version of a series regulator is that of a resistor in series with a zener diode. The problem with this scheme is the relatively low range of output

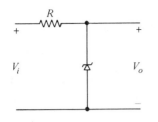

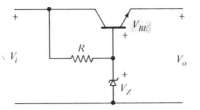

FIGURE 3.1 SIMPLE ZENER VOL-
TAGE REGULATOR

FIGURE 3.2 SIMPLE SERIES VOL-
TAGE REGULATOR

current. This simple series regulator is shown in Fig. 3.1. Another simple version
improves the output current range by adding a power transistor to provide the
needed current gain, as shown in Fig. 3.2. Note that this regulator must have a
load in order for the transistor to be on. The regulated output voltage for this
circuit is $V_Z - V_{BE}$.

From these two simple circuits we can see some requirements for all series
regulators. First, the input voltage must be larger than the output. Secondly, the
lower the output impedance, the better the regulator.

Before we take up regulator characteristics, consider the block diagram in
Fig. 3.3, where an improved series regulator of the type used in IC regulators is
outlined. The unregulated input is applied to generate a reference voltage and to
bias the error amplifier. The output voltage sample is compared with the reference voltage by the error amplifier. The error signal is then applied to the

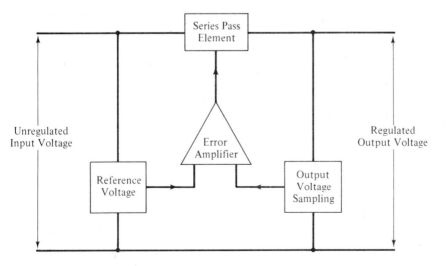

FIGURE 3.3   SERIES REGULATOR BLOCK DIAGRAM

series-pass element, typically an *NPN* power transistor. For varying input as well as load conditions, the reference voltage ideally remains constant. Thus should the output voltage be low, the error signal is amplified by the error amplifier, causing the series-pass element to increase output current and therefore restore the proper output voltage. Conversely, should the output voltage be too high, the error signal, now of opposite polarity, is again amplified by the error amplifier, causing the series-pass element to decrease the output current. This once again restores the proper output voltage.

Although the specific circuit of a series regulator may take on one of many versions, the basic operation outlined above remains essentially unchanged.

3.2   VOLTAGE REGULATOR PARAMETERS

The characteristics of an ideal voltage regulator are: no output voltage change due to load or input variations, infinite ripple rejection, zero output impedance, zero output noise over all frequencies, and no temperature dependence of any parameters. However, ideal voltage regulators do not exist, so let us examine some of the more common parameters and their definitions as they pertain to realizable regulators.

LOAD REGULATION   The change in output voltage for a change in load current. This quantity is usually measured and specified as the load current is changed from essentially zero (or some low value) to its maximum value, resulting in an output voltage change. Sometimes, the percentage of change in the output voltage for this change in load current is specified rather than the output voltage limits.

LINE REGULATION   The change in output voltage for a specific change in input voltage. This quantity may be specified as a percentage: for example a 0.1% load regulation figure means that a 1 V change in input voltage causes the output voltage to change by 1 mV. Alternately, the total change in output voltage may be specified for a change in input voltage from its minimum-to-maximum value variation. For example, an output voltage change of 200 mV for an input voltage change from 7 to 27 V is specifying the same load regulation as the 0.1% figure.

RIPPLE REJECTION   The ratio of the input (rms) ripple voltage to the output (rms) ripple voltage, usually expressed in dB. For example, a ripple rejection of 80 dB indicates that a 1 V ripple at the input produces only a 0.1 mV ripple at the output. (This parameter is quite similar to the PSRR for OP AMPs.)

OUTPUT IMPEDANCE   The effective impedance that appears in series with the output. The real significance is that its value increases with frequency; the dc performance of the regulator at different currents is adequately prescribed by the load regulation, while the output impedance is an indication of how well the regulator maintains the stated output voltage when subjected to a load requiring the load current to change rapidly. (An example of such a load is any digital circuit.)

DROPOUT VOLTAGE   The minimum differential between the input and output voltages required to keep the regulator operating. This parameter may also be stated as a minimum input voltage for a given output voltage, but the essential characteristic is that the input voltage must under all conditions be sufficiently higher than the desired output to keep the regulator on.

OUTPUT NOISE VOLTAGE   That portion of the output noise or ripple voltage that is generated by the regulator itself. This quantity is measured with no input ripple and under constant load conditions.

LONG-TERM STABILITY   The stability of the output voltage over 1000 hrs. It is usually specified as a percentage of the output voltage per 1000 hrs. For example, a regulator may have a drift of less than 0.1% in the output voltage over a 1000 hr period.

QUIESCENT OR STAND-BY CURRENT   The bias current required by the regulator to provide the stated output voltage at zero load current. While the bias current of the regulator does increase with increasing load current, the input current required at a given load current must be greater than the stated load current by an amount specified by the quiescent current specification.

POWER DISSIPATION   The maximum power that can safely be dissipated in the regulator under any load or input conditions. This quantity determines the combination of safe load current for a particular input voltage (or vice versa) at a given output voltage. Essentially, the power dissipated in the regulator is the load current times the input-output voltage differential at that current. Note that the worst-case of input-output voltage differential occurs when the input voltage is highest (due to line fluctuations) and the output voltage is lowest.

TEMPERATURE COEFFICIENT   The average change in output voltage for a 1°C change in regulator temperature, usually specified in percent/°C. This quantity is especially significant for IC voltage regulators, since at high output currents the higher internally dissipated power forces the junction temperature of all devices to rise and tends to change the output voltage.

3.3 REGULATOR CIRCUITS

An example of a series regulator is shown in Fig. 3.4. A zener diode is used to provide the reference voltage; resistors R_1 and R_2 provide output voltage sampling; the OP AMP acts as a high-gain error amplifier driving an *NPN*

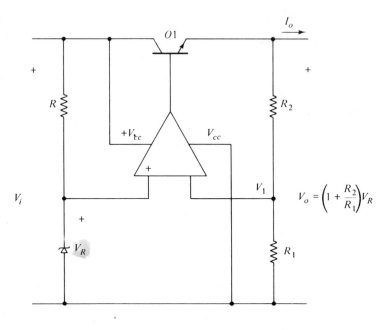

FIGURE 3.4   SERIES REGULATOR SCHEMATIC $(V_o > V_Z)$

series-pass transistor, $Q1$. The OP AMP is operated in a non-inverting mode with the gain set by R_1 and R_2 and a constant input of V_R. Thus the regulated output voltage is:

$$V_o = \left(1 + \frac{R_2}{R_1}\right)V_R$$

If we assume the OP AMP power supply rejection to the ideal, then the load regulation of the circuit is determined by the regulation of the zener. If R_Z is the zener dynamic resistance, then:

$$\text{Load Regulation} = \left(\frac{R_Z}{R + R_Z}\right)\left(\frac{R_1 + R_2}{R_1}\right)$$

As an example, consider V_i to be between 15 and 20 V, $R = 1$ k, and $V_Z = 6.8$ V with $R_Z = 25\ \Omega$. For a nominal 10 V output, the ratio of R_2 to R_1 should be 0.47. Typically we might choose $R_1 = 10$ k and $R_2 = 4.7$ k. For this example, the load regulation would be 0.036. This would keep the output voltage change below

0.18 V for the stated variation in input voltage of 10 V. To see how this is accomplished, consider that with a nominal input of 17.5 V, the output voltage is exactly 10 V, which OP AMP action requires to equalize the voltages at its inverting and non-inverting inputs. Thus the voltage across R_1 is equal to V_R. With no load, this requires a current of approximately 0.68 mA through $Q1$ (and R_1 and R_2). If the input voltage should decrease, the current through $Q1$ would also decrease (even though only momentarily). This would decrease V_1, causing a positive error voltage at the input of the OP AMP. Thus the output of the OP AMP would swing more positive, turning on $Q1$ more, and the emitter current of $Q1$ would increase to bring the voltage across R_1 back into equilibrium with V_R. This action re-establishes the output voltage.

For an increase in input voltage, $Q1$ increases the current through R_1, causing a negative error voltage at the input of the OP AMP. The OP AMP output now swings in the negative direction causing $Q1$ to conduct less until its emitter current is decreased sufficiently to equalize the voltage across R_1 with V_R. This again re-establishes the original output voltage.

For regulators where the desired output voltage is less than the reference voltage, the circuit shown in Fig. 3.5 can be used. Here, the OP AMP is

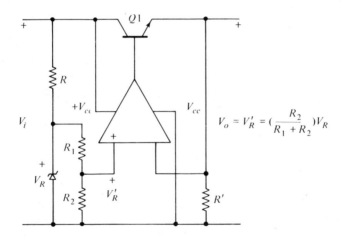

$$V_o = V_R' = \left(\frac{R_2}{R_1 + R_2}\right)V_R$$

FIGURE 3.5   SERIES REGULATOR FOR $V_o < V_Z$

connected as a voltage follower, with an input voltage derived from a voltage divider across the zener. The output voltage then must equal V_R':

$$V_o = \left(\frac{R_2}{R_1 + R_2}\right)V_R$$

The load regulation for this circuit is:

$$\text{Load Regulation} = \left(\frac{R_Z}{R + R_Z}\right)\left(\frac{R_2}{R_1 + R_2}\right)$$

where R_Z is assumed significantly smaller than either R_1 or R_2, and the OP AMP PSRR is assumed negligibly large.

As an example, consider V_i to be again in the range of 15 to 20 V and $V_Z = 6.8$ V with $R_Z = 25\ \Omega$; the desired output voltage is 5 V. The ratio of R_2 to R_1 should be 0.36 (we might choose $R_1 = 3.3$ k and use a 10 k potentiometer for R_2 trimmed to 9.17 k). The load regulation for this circuit is then 0.0065. Thus for a 5 V change in the input, we would expect the output to change by about 32 mV. The resistor R' is included as a bleeder for $Q1$ to insure that $Q1$ conducts with no load, and can be omitted when the regulator is driving a load.

A comparison of the two regulators shown in Figs. 3.4 and 3.5 indicates that the load regulation for the circuit in Fig. 3.5 is better. Nevertheless, this circuit is not necessarily the more desirable since it requires a larger input-output voltage differential and thus gives higher power dissipation in the series-pass transistor at the same load current. If the load current in the two examples is 0.5 A, the power dissipated in $Q1$ is 5 W for the circuit in Fig. 3.4 and 7.5 W for the circuit in Fig. 3.5 (calculated at highest input voltage). Both regulators have a limit on the maximum load current imposed by the OP AMP. If a 741 is used, its output current is limited to approximately 25 mA. Since this is the base current of $Q1$, the load current is limited to $(\beta + 1)$ times this current. If a 2N3055 silicon NPN transistor is used for $Q1$ with a minimum β of 40, a maximum load current of approximately 1 A can be expected.

From the previous discussion, it should be obvious that the regulation is limited to that of the reference. A significant improvement in regulation is possible for both regulator circuits if the zener current is regulated by a constant current source. This change is implemented by replacing R with an FET or a transistor current source, as shown in Fig. 3.6. The zener current in the FET circuit is I_{DSS}, and the FET is chosen accordingly. The input voltage must be higher than the sum of the zener voltage and the FET pinch-off voltage to insure constant current. To determine the load regulation, use the FET r_{ds} instead of R in the equations above. For the transistor circuit, the zener current is set by $(V_{Z1} - V_{BE})/R_E$. Although this version allows for higher and more easily adjustable zener currents, it typically requires higher minimum input voltages: V_i should be larger than $1 + V_R + V_{Z1}$ (the 1 volt being necessary to keep the transistor in its active region). The regulation for this version is determined by substituting $1/h_{oe}$ for R in the above equations.

Negative voltage regulators can be implemented with duals of the circuits in Figs. 3.4 and 3.5: the NPN series-pass transistor is replaced by a PNP, the OP

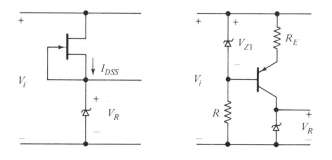

FIGURE 3.6   ZENER CURRENT REGULATION

AMP supply leads are reversed, as is the zener, and the polarities of V_i and V_o are reversed. The operation of these negative regulators is identical in principle to that of the positive regulators already discussed.

3.4   CURRENT-LIMITING

Although the regulators described in Figs. 3.4 and 3.5 do have current-limiting provided by the OP AMP, in most cases a programmable current-limiting scheme is necessary to protect the regulator and/or the load. Such a scheme is easily implemented, as shown in Fig. 3.7, by the addition of a resistor and a transistor. As long as the load current is less than V_{BE}/R_{SC}, transistor $Q2$ is essentially off, and the regulator performance is the same as if $Q2$ and R_{SC} were not in the circuit. However, when the load current is sufficient to set up approximately 0.7 V across R_{SC}, $Q2$ is biased on and conducts. Now as the output voltage falls below the set value and I_{error} increases to raise it, $Q2$ siphons off this increase (as I_{c2}), and I_{B1} does not increase, but remains essentially constant.

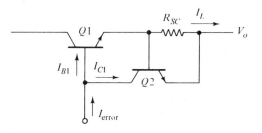

FIGURE 3.7   CURRENT-LIMITING

Thus the load current (which is essentially equal to βI_{B1}) is forced to stay constant. As the load resistance decreases, requiring more load current to maintain the output voltage, the output voltage decreases proportionally since the current is fixed. The regulator is now maintaining constant load current instead of constant output voltage. Note that the current is kept constant even if the output is short-circuited. Thus:

$$I_{SC} = \frac{V_{BE}}{R_{SC}} = \frac{0.7 \text{ V}}{R_{SC}}$$

The regulator characteristics with this form of current-limiting are shown in Fig. 3.8. There are two distinct modes of operation: first, the conventional voltage regulator mode where the output voltage is essentially constant; then the second mode where the load current is essentially constant.

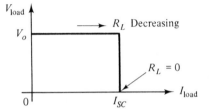

FIGURE 3.8   REGULATOR CHARACTERISTICS WITH CURRENT-LIMITING

In a series regulator, the power dissipation in the SPE ($Q1$) is a minimum when the load current is a minimum; similarly, as the load current is increased, so is the power dissipated in the SPE. Obviously, the highest power dissipated in the SPE occurs when the output is short-circuited to ground. Then we have:

$$P = V_i I_{SC}$$

We can improve on the power dissipation without sacrificing the maximum output current with the *foldback current-limiting* scheme shown in Fig. 3.9. Here, while the load current sensing is still accomplished by R_{SC}, there is a voltage divider formed by R_a and R_b to the base of $Q2$. Consider the load resistance to be decreasing, thus increasing the load current. As long as the base-emitter voltage of $Q2$ is insufficient to turn it on, the output voltage is regulated. When the voltage (due to I_L) across R_{SC} increases to the point where $Q2$ conducts, we have:

$$V_{Rb} = V_o + V_{BE} = V_{L'}\left(\frac{R_b}{R_a + R_b}\right)$$

Solving for $V_{L'}$ above and noting that $V_{L'} = I_L R_{SC} + V_o$, we get:

$$I_L = \frac{V_o R_a}{R_{SC} R_b} + \frac{V_{BE}}{R_{SC}}\left(1 + \frac{R_a}{R_b}\right) = I_{knee}$$

When $Q2$ conducts, current limiting starts. At this point, output current, called the knee current, is a function of the output voltage as well as the programming resistors, R_a, R_b, and R_{SC}.

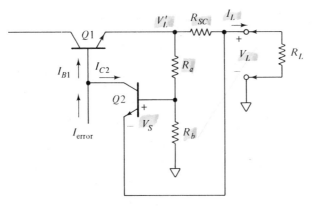

FIGURE 3.9   FOLDBACK CURRENT-LIMITING

If the output is shorted to ground, the output voltage is zero; thus the load current is:

$$I_{SC} = \frac{V_{BE}}{R_{SC}}\left(1 + \frac{R_a}{R_b}\right)$$

Note that I_{knee} is larger than I_{SC}; in fact:

$$I_{knee} = I_{SC} + \frac{V_o R_a}{R_{SC} R_b}$$

Thus the maximum current at the rate output voltage for the regulator (this is

I_{knee}) is larger than the short-circuit current. These characteristics are shown in Fig. 3.10. There is a tremendous advantage in having the power supply protected against shorts that may be accidental or due to component failure of the circuit being supplied.

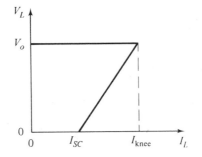

FIGURE 3.10   FOLDBACK CURRENT-LIMITING CHARACTERISTICS

In general, the maximum power dissipation of the SPE determines both the short-circuit and the knee currents:

$$I_{SC} \leq \frac{P_{\text{max}}}{V_{i\text{max}}}$$

and

$$I_{\text{knee}} \leq \frac{P_{\text{max}}}{V_{i\text{max}} - V_o}$$

Once the output voltage and the two currents are known, the above equations can be rearranged and solved for the resistors needed to satisfy the requirements:

$$\frac{R_b}{R_a} = \left[\left(\frac{V_o}{V_{BE}}\right)\left(\frac{I_{SC}}{I_{\text{knee}} - I_{SC}}\right)\right] - 1$$

and

$$R_{SC} = \left(\frac{R_a}{R_b} + 1\right)\frac{V_{BE}}{I_{SC}}$$

A ready comparison between the conventional and foldback current-limiting schemes can be made by using the following example. Consider a raw dc input of 15 V and the regulator of Fig. 3.4 adjusted for an output of 10 V. The power dissipation for $Q1$ is to be less than 5 W (worst case). Obviously the worst case is when the output is short-circuited, thus the short-circuit current should be 5/15 A or less. Assuming $V_{BE} = 0.7$ V, this gives us $R_{SC} = 2.1\Omega$ for the conventional current-limiting circuit in Fig. 3.7. If we design the foldback current-limiting circuit of Fig. 3.9, the knee current should be less than 5/(15-10) A = 1 A; the short-circuit current is the same as previously. Using these values, we determine $R_b/R_a = 6.14$ and $R_{SC} = 2.4\Omega$. (We might use $R_a = 330\Omega$ and $R_b = 2$ k.) The difference between the two current-limiting schemes is that the maximum current at the specified voltage is always higher for the foldback scheme than for the conventional current-limiting (in the example, 1 A as opposed to 333 mA).

Thus the purpose of foldback current-limiting is to extend the output current while maintaining the same power dissipation, or to lower the power dissipation at the same output current.

Another technique that is useful in many applications is simply to shut down the regulator should a faulty condition be present. This is done by starving the base of the SPE ($Q1$) as shown in Fig. 3.11. The circuit shows two remote inputs, V_1 and V_2. If either of these inputs (or both) increases in voltage, $Q2$ turns on and pulls the error amplifier output low, thus effectively turning off $Q1$. The fault condition (excess output current, output shorted, component failure, etc.) is sensed and the condition is stored in a latch (see sections on digital ICs), whose output is applied to $Q2$ through R. Once the fault has been removed, the latch is reset (manually or automatically), $Q2$ turns off (if *all* the fault conditions are

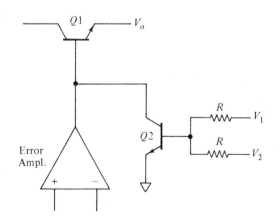

FIGURE 3.11   REMOTE SHUTDOWN OF REGULATOR

removed), and the regulator re-establishes the proper output voltage. Note that this scheme can be extended to shut down the regulator for any number of fault conditions simply by inserting additional resistors R to the base of $Q2$.

3.5   TYPE 723 REGULATOR

As an example of a monolithic voltage regulator, consider the type 723 regulator whose block diagram is shown in Fig. 3.12. On a single monolithic chip we have available the building block to construct a high quality regulator for positive or negative outputs, series type or switching, with or without current limiting, and with remote shutdown capability.

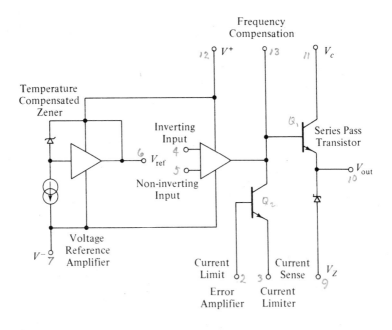

FIGURE 3.12   723 REGULATOR BLOCK DIAGRAM

Before examining all these features, let us consider the basic series regulator as implemented, using the 723. Note that since the reference voltage is nominally 7 V (see data sheets in Appendix), two basic regulator configurations for positive voltage applications result: one for V_o greater than V_R, one for V_o less than V_R. The two cases are shown in Figs. 3.13 and 3.14, respectively, both utilizing conventional current-limiting.

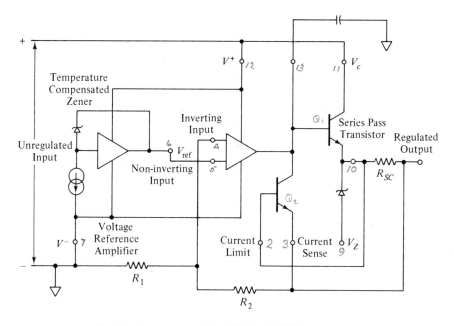

FIGURE 3.13   723 POSTIVE REGULATOR ($2 \leqslant V_o \leqslant 7V$)

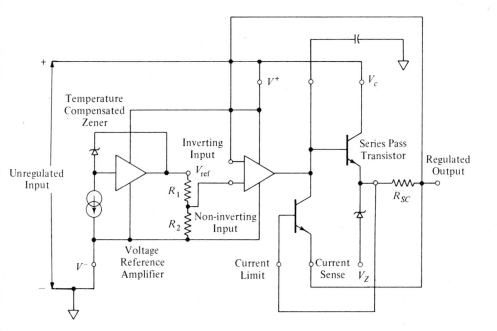

FIGURE 3.14   723 POSITIVE REGULATOR ($7 \leqslant V_o \leqslant 37V$)

To apply the 723 in any circuit configuration, the maximum device specifications must be observed. For example, the maximum current that may be drawn from V_R is 15 mA. This means that the minimum resistance that can be connected from V_R to $V-$ is 7/15 k, or approximately 500Ω. Similarly, the reverse breakdown voltage for the transistors in the 723 is in excess of 40 V. So the maximum differential between $V+$ and $V-$ must be maintained below 38 V. The total device dissipation is to be limited to 800 mW or less; thus the difference between the input and output voltages times the output current should be less than 800 mW.

Two additional conditions must be met for proper operation: the error amplifier inputs must be at least 2 V above $V-$ in order for the transistors in the error amplifier to be biased; secondly, the minimum input voltage must be at least 3 V higher than the output voltage, or a minimum of 9.5 V, whichever is higher. The 3 V minimum input-output differential is to make certain that the SPE does not saturate under heavy output current conditions. The absolute 9.5 V minimum is to bias the 7 V zener and its associated current regulator and buffer.

To see how the 723 can be used, consider the following example: the input available is 18 Vdc with a maximum 4 V peak-to-peak ripple for currents equal to or less than 100 mA. The desired output is 9 V at 60 mA with short-circuit protection.

The first step is to check if the 723 is compatible with the input and output requirements. The input is between $18 - 2 = 16$ V and $18 + 2 = 20$ V; thus it is well within the specified range for the 723. The minimum and maximum input-output voltage differentials are 7 V and 11 V respectively; these values are well within the specifications of 3 and 38 V. At the rated output of 9 V at 60 mA, the power dissipation in the 723 is $(18 - 9)60$ mW = 540 mW. This is lower than the 800 mW maximum for the 723, so no external power transistor is necessary. In order to maintain the maximum power dissipation of 800 mW or less for an 18 V input, the short-circuit current should be 44 mA or less. Since this value is lower than the desired output current, foldback current-limiting must be used. If we choose to keep the power dissipation under short-circuit conditions the same as for maximum output current at the rated voltage (540 mW), we would then design for a short-circuit current of 540/18 mA = 30 mA. We next choose the knee current equal to the desired maximum output current of 60 mA. We could now calculate the resistors necessary by assuming V_{BE} = 0.7 V and using the relationships developed in the previous section. However, with 540 mW dissipated, the junction temperature may be well in excess of 25°C, so we should calculate V_{BE} somewhat more accurately. As a rule, V_{BE} decreases approximately 2 mV for each °C rise in temperature. The junction temperature at a given power dissipation is obtained:

$$T_J = 25 + \Theta_{JA}P = 25 + (111)(0.54) \cong 85°C$$

where Θ_{JA} is the thermal resistance of the 723, listed in the data sheet to be $111°C/W$. Thus:

$$V_{BE} = 0.66 - (0.002)(T_J - 25) = 0.54 \text{ V}$$

where the ambient ($25°C$) value of $V_{BE} = 0.66$ is obtained from the data sheet. We now calculate the resistors for the foldback circuit:

$$\frac{R_b}{R_a} = \left(\frac{9}{0.54}\right)\left(\frac{30}{60 - 30}\right) - 1 = 15.7$$

We might choose $R_a = 330\Omega$, then $R_b = (15.7)(330) \cong 5.1$ k. Next, calculate R_{SC}:

$$R_{SC} = \left(\frac{0.33 + 5.1}{5.1}\right)\left(\frac{0.54}{0.03}\right) = 19\Omega$$

The design is finished by calculating R_1 and R_2: $V_o = (1 + R_2/R_1)V_R$, so:

$$\frac{R_2}{R_1} = \frac{V_o}{V_R} - 1 = \frac{9}{7} - 1 \cong 0.286$$

where we have used the nominal value of 7 V for V_R. If we choose $R_1 = 15$ k, then $R_2 = (15)(0.286) = 4.3$ k. To allow for some uncertainty in V_R, we might use a 5 k potentiometer for R_2 and trim for the exact output voltage. The complete circuit and $V_o - I_o$ characteristics for this example are shown in Fig. 3.15.

As mentioned earlier, the 723 can be used to provide regulation for negative voltages, as shown in Fig. 3.16. In this circuit, the 723 output is level-shifted by the zener in series with the output. The external zener maintains a 12 V differential across the 723. The regulated output is stepped down by the voltage divider formed by R_1 and R_2 to provide the non-inverting input to the error amplifier. Note that due to the connection of $V-$ to the output, the voltage at V_R with respect to ground is $V_o - V_R$. With $R_3 = R_4$, the inverting input of the error amplifier is 3.5 V ($V_R/2$) below V_o. At equilibrium the voltages at the inverting and non-inverting inputs of the error amplifier are equal. Thus the voltage across R_1 must also be 3.5 V. Since the same current flows through R_2 as through R_1, the output voltage is:

$$V_o = -3.5\left(1 + \frac{R_2}{R_1}\right)$$

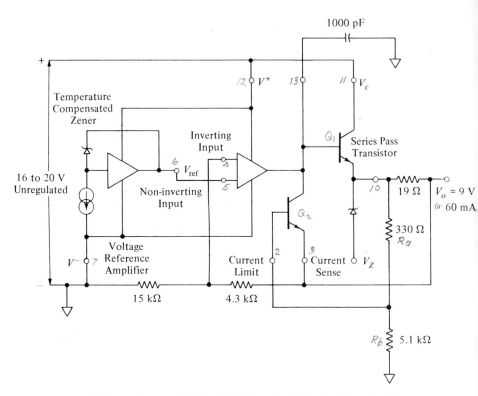

FIGURE 3.15   723 REGULATOR DESIGN CIRCUIT (see text)

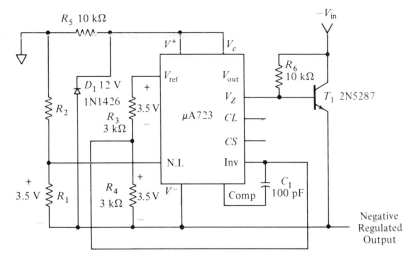

FIGURE 3.16   NEGATIVE FLOATING REGULATOR

Since the output of the error amplifier is fed through the zener and the maximum zener current is 25 mA, an external series-pass transistor is used to boost the output current. Note that the regulator is self-starting since the pass transistor is biased by R_6. This is necessary for this configuration since the bias for the 723 is not derived from the unregulated input, but is developed from the output. Also note that the addition of D_1 and R_5 allows the regulator output to be well in excess of the 38 V maximum for the 723. If $R_1 = 3.5$ k and $R_2 = 96.5$ k, a negative 100 V regulator results.

The same technique of "floating" the 723 can be used to provide positive voltage regulators for outputs in excess of the 38 V maximum for the 723. The only limitation is the breakdown voltage of the external series-pass transistor. (In the positive regulator, the addition of an external pass transistor forms a Darlington configuration and only boosts the maximum output current without changing any other aspects of the basic regulator.) Consider the positive floating regulator circuit shown in Fig. 3.17. The high input voltage is reduced by the 12 V zener so that the voltage differential between $V-$ and $V+$ is only 12 V. Note the duality of this configuration and that of Fig. 3.16: the voltage across R_3 and R_4 is $V_R/2$ or 3.5 V. At equilibrium the voltage across R_1 must be the same as that

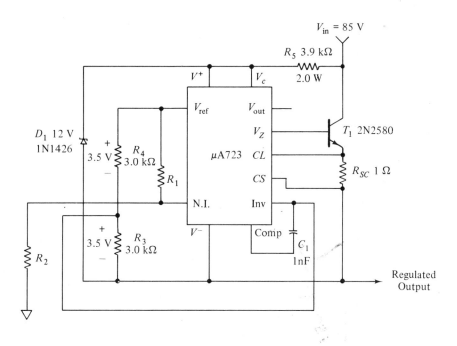

FIGURE 3.17   POSITIVE FLOATING REGULATOR

across R_4, since the inverting and non-inverting inputs of the error amplifier are at the same voltage. The output voltage is then:

$$V_o = 3.5 \left(1 + \frac{R_2}{R_1} \right)$$

For example, if $R_1 = 3.5$ k and $R_2 = 46.5$ k, the output voltage is 50 V.

We need to note that the error amplifier in the 723 does not have internal compensation, and that an external capacitor is needed to prevent oscillation.

The 723 regulator can also be used as a switching regulator (see data sheets in Appendix). Switching regulators are used where power efficiency (percent of input power delivered to the load) is critical. Such applications are relatively rare—for extremely large output currents when the input voltage is much larger than the desired output voltage. The basic idea in switching regulators is controlling the duty cycle of an oscillator whose output is rectified and filtered. Since the transistors in the oscillator are either saturated (high current, low voltage) or off (high voltage, no current), the power dissipated in the regulator is minimized. Thus most of the power drawn from the input is delivered to the output.

The 723 regulator can also be used in the remote shutdown mode as shown in Fig. 3.18. Operation is the same as that of Fig. 3.11 discussed previously. However, if conventional current-limiting is not used, the current-limit transistor can be used for the remote shutdown function.

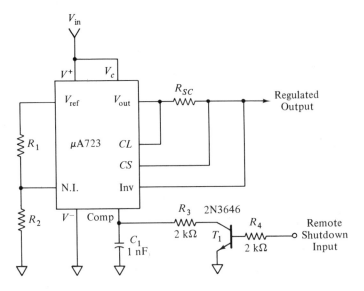

FIGURE 3.18   REMOTE SHUTDOWN REGULATOR

Additional applications for the 723 regulator can be found in the manufacturers' data sheets and application notes; representative data sheets are in the
Appendix.

3.6   THREE-TERMINAL REGULATORS

While adjustable regulators like the 723 offer tremendous flexibility and
many varied configurations, the fixed-voltage three-terminal regulators offer
excellent performance and the utmost in simplicity for the majority of power
supply needs. Basically, the three-terminal regulator IC offers a single, three-
lead device that contains the error amplifier, reference voltage, series-pass element, and the circuit to set the output voltage. In order to use the regulator, we
merely connect the input, common, and output terminals.

Table 3.1 gives a listing of the commonly available three-terminal regulators. Notice that almost all voltages between $\pm$ 5 and ± 24 V are available.
There are basically three package types: K is the same as the TO-3 power
transistor case, H is the same as the TO-5 transistor case, and T is similar to the
plastic power transistor case TO-220. The three different packages correspond to
different output current maximums.

Three-terminal regulators all contain internal current-limiting. However,
note that since the current-limiting transistor is on the same chip as the series-
pass element, it is subjected to the same temperature. As the power dissipated in
the SPE increases with increasing output current, so does the junction temperature of both the SPE and the current-limiting transistor. Thus the limiting transistor begins to turn on at a lower V_{BE}. The output current limit then decreases as
the dissipation in the regulator increases. In essence, we no longer have simple
current-limiting—we have power-limiting. This feature is only possible on an
integrated circuit due to the nature of the thermal coupling between the SPE and
limiting transistor. If we attempted the same operation in discrete form by physically attaching the limiting transistor to the power SPE, the long thermal time
delay between the heating of the power transistor and the subsequent heating of
the limiting transistor would defeat the usefulness of the negative thermal feedback.

To supply the large output current (3 A for the 323), the SPE in the three-
terminal regulator may take up more than half the total chip area: it is a multiple
emitter transistor (MET) with as many as 30 or 40 emitters. This is necessary in
order to lower the current density through the emitter.

In the application notes for the specific regulator to be used, the recommended external compensation (if any) is given. Although the internal error
amplifier is frequency compensated, any lead inductance between the rectifier

TABLE 3.1   THREE TERMINAL REGULATORS

TYPE NO.	OUTPUT VOLTAGE	OUTPUT CURRENT*
309 H	+5	200 mA
309 K	+5	1 A
320 H	−5, −5.2, −6, −8, −12, −15, −18, −24	200 mA
320 K or T	−5, −5.2, −6, −8, −12, −15, −18, −24	1 A
323 K	+5	3 A
340 K or T	5, 5.2, 6, 8, 12 15, 18, 24	1 A
341 P	5, 6, 8, 12, 15, 18, 24	500 mA
342 H or P	5, 6, 8, 10, 12, 15, 18, 24	200 mA
345	−5	3 A
7805 K or T	5	1 A
7806 K or T	6	1 A
7808 K or T	8	1 A
7812	12	1 A
7815	15	1 A
7818	18	1 A
7824	24	1 A

Approximate values. Devices are temperature limited.

filter capacitor and the input of the regulator would cause the regulator to oscillate. A similar situation exists at the output: any inductive load component needs to be neutralized by a bypass capacitor. A typical application of a three-terminal regulator is shown in Fig. 3.19.

Let us examine the 309 fixed 5 V regulator since it is quite representative and probably the most commonly used (TTL applications). The electrical characteristics are listed in Table 3.2. One additional parameter is important: the max-

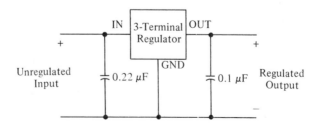

FIGURE 3.19   TYPICAL THREE-TERMINAL REGULATOR APPLICATION

TABLE 3.2 5 V REGULATOR ELECTRICAL CHARACTERISTICS (Note 1)

PARAMETER	CONDITIONS	LM109			LM309			UNITS
		MIN	TYP	MAX	MIN	TYP	MAX	
Output Voltage	$T_j = 25°C$	4.7	5.05	5.3	4.8	5.05	5.2	V
Line Regulation	$T_j = 25°C$ $7V \leqslant V_{IN} \leqslant 25V$		4	50		4	50	mV
Load Regulation	$T_j = 25°C$							
TO-5	$5mA \leqslant I_{OUT} \leqslant 0.5A$		20	50		20	50	mV
TO-3	$5mA \leqslant I_{OUT} \leqslant 1.5A$		50	100		50	100	mV
Output Voltage	$7V \leqslant V_{IN} \leqslant 25V$ $5mA \leqslant I_{OUT} \leqslant I_{max}$ $P < P_{max}$	4.6		5.4	4.75		5.25	V
Quiescent Current	$7V \leqslant V_{IN} \leqslant 25V$		5.2	10		5.2	10	mA
Quiescent Current Change	$7V \leqslant V_{IN} \leqslant 25V$			0.5			0.5	mA
	$5mA \leqslant I_{OUT} \leqslant I_{max}$			0.8			0.8	mA
Output Noise Voltage	$T_A = 25°C$ $10Hz \leqslant f \leqslant 100\ kHz$		40			40		μV
Long Term Stability				10			20	mV
Thermal Resistance Junction to Case (Note 2)								
TO-5			15			15		°C/W
TO-3			3			3		°C/W

NOTES:

1. Unless otherwise specified, these specifications apply for $-55°C \leqslant T_j \leqslant 150°C$ for the 109 or $0°C \leqslant T_j \leqslant 125°C$ for the 309. $V_{IN} = 10V$ and $I_{OUT} = 0.1A$ for the TO-5 package or $I_{OUT} = 0.5A$ for the TO-3 package. For the TO-5 package, $I_{max} = 0.2A$ and $P_{max} = 2.0W$. For the TO-3 package, $I_{max} = 1.0A$ and $P_{max} = 20W$.

2. Without a heat sink, the thermal resistance of the TO-5 package is about 150°C/W, while that of the TO-3 package is approximately 35°C/W. With a heat sink, the effective thermal resistance can only approach the values specified, depending on the efficiency of the sink.

imum input voltage is 35 V. We digress briefly to note that the 109 and 309 are identical circuits, the only difference being that the 109 characteristics are guaranteed over the full military temperature range from −55°C to 125°C, while the 309 is the commercial version whose characteristics are guaranteed from 0°C to 70°C. In examining the electrical characteristics, note the excellent line and load regulation, as well as the low output noise voltage. As mentioned in Table 3.1, the output current is not limited in a conventional manner. As we can see in Fig. 3.20, the higher the input voltage and thus the power dissipated in the regulator, the lower the peak output current for both the H and K packages. The maximum power dissipation as a function of room temperature is given in Fig. 3.21. Note the higher power-handling capabilities if a suitable heat sink is used.

K PACKAGE (TO-3)

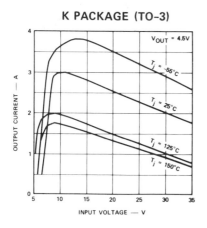

H PACKAGE (TO-5)

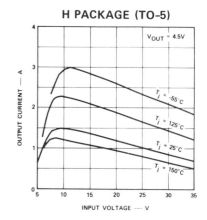

FIGURE 3.20   309 PEAK OUTPUT CURRENT

LM309 (TO-5)

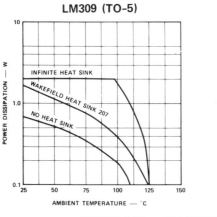

LM309 (TO-3)

FIGURE 3.21   309 MAXIMUM POWER DISSIPATION

The minimum input voltage for the rated output as a function of load current and junction temperature is shown in Fig. 3.22. Note that if the input is above 7 V, the regulator is properly biased under all junction temperature and load current conditions.

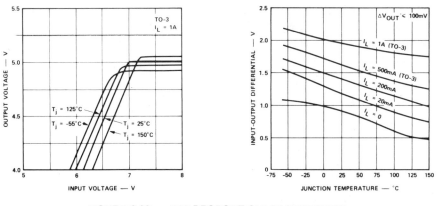

FIGURE 3.22   309 DROPOUT CHARACTERISTICS

The regulator output impedance increases significantly for frequencies above 10 kHz, as indicated in Fig. 3.23. Thus for high-speed switching applications (which is the normal case for the 309 used to supply TTL digital circuits), it is advisable to bypass the output with a 0.1 μF or larger disc capacitor.

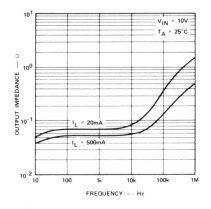

FIGURE 3.23   309 OUTPUT IMPEDANCE

The quiescent or bias current for the regulator as a function of temperature for two values of load current is given in Fig. 3.24. The important thing to note from this figure is the typical bias current of approximately 5 mA. We shall make use of this value later.

There are only a few key requirements in applying the 309 regulator for a fixed 5 V supply: make certain that the input dc voltage *minus the peak ripple voltage* exceeds the 7 V minimum; use an appropriate heat sink to insure that the maximum power dissipation is not exceeded; and lastly, use more than one regulator if the circuit being supplied is segmented on a number of printed circuit boards. In fact it is a good idea to use one regulator per card, remembering to include the input bypass capacitor.

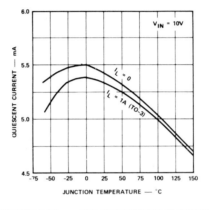

FIGURE 3.24   309 QUIESCENT CURRENT

Although the 309 and other three-terminal regulators like it are primarily intended to supply the rated voltage, they can be used to supply higher voltages as shown in Figs. 3.25 and 3.26. In Fig. 3.25, the 309 output sets the voltage across R_1 to 5 V, thus $I_1 = 5/R_1$. The output voltage is then 5 V plus the voltage

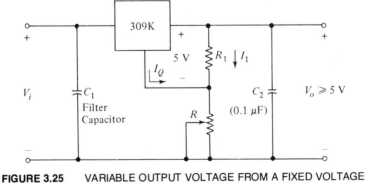

FIGURE 3.25   VARIABLE OUTPUT VOLTAGE FROM A FIXED VOLTAGE
REGULATOR

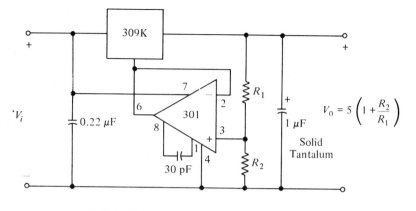

FIGURE 3.26   ADJUSTABLE REGULATOR

across R_2 which is $(I_Q + I_1)R$. With a bias current of 5 mA and if R_1 is 330Ω, then $I_1 \cong 15$ mA. The output voltage is:

$$V_o = 5 + 0.02R$$

For example, if R is a 1 k potentiometer, the output voltage is adjustable between a minimum of 5 V and a maximum of 25 V.

For variable output voltage applications, the 317 regulator offers improved performance over the 309. It has lower bias current, and allows the output voltage to be varied from a low of 1.2 V to a maximum of 37 V.

The circuit of Fig. 3.26 removes the dependence of the output voltage on the 309 regulator bias current. In essence, the 309 is being used as a series-pass element. The 301 OP AMP is in a voltage follower configuration; the inverting input is at the output voltage minus the 5 V forced by the 309. At equilibrium, the non-inverting input is also 5 V below the output voltage, thus forcing 5 V across R_1 which sets the current through R_2. The output voltage is then:

$$V_o = 5 \left(1 + \frac{R_2}{R_1} \right)$$

The purpose of the 301 OP AMP in this circuit is clearly to buffer R_2 from the regulator bias current while sinking the same bias current at its output.

In applications where the available unregulated input voltage exceeds the maximum allowable input for the regulator, a pre-regulator is used as shown in Fig. 3.27. Note that $Q1$ must pass the same output current as the three-terminal regulator, and therefore should have the appropriate current rating.

A simple modification of the basic three-terminal regulator circuit provides a current regulator, as shown in Fig. 3.28. The regulator sets up its rated output

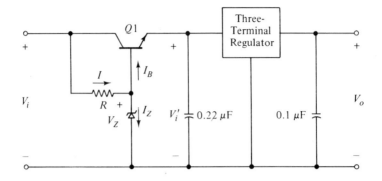

FIGURE 3.27   LOWERING INPUT VOLTAGES WHICH ARE TOO HIGH

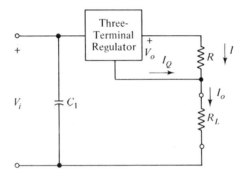

FIGURE 3.28   THREE-TERMINAL VOLTAGE REGULATOR USED AS A
CURRENT REGULATOR

voltage across R, so that $I = V_o/R$. The output current is:

$$I_o = I_Q + V_o/R$$

For example, if a 309 is used with $R = 20\ \Omega$, the load current is constant at
approximately 250 mA. For an input voltage of 20 V, the output current regulation is maintained for any R_L between zero and 52 Ω (the voltage across the load
must be less than the input by the minimum input-output differential plus V_o or,
in our example, 7 V).

A simple dual (plus and minus) voltage regulator, useful for powering OP
AMP circuits, is shown in Fig. 3.29. It utilizes two 3-terminal regulators, the
340T-15 for the +15 V output and the 320T-15 for the −15 V output. The two
diodes across the outputs allow the regulators to start under a common load and
should be capable of sustaining the regulator short-circuit current. Note that the
negative regulator requires a bypass capacitor across the output for stability. This
should be a low inductance type, solid tantalum capacitor, for example. If an
aluminum electrolytic is used, the value should be increased to at least 25 μF.

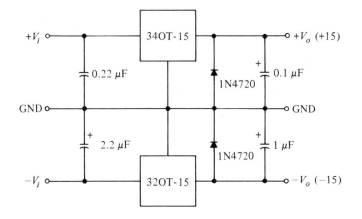

FIGURE 3.29   DUAL POWER SUPPLY (SOLID TANTALUM CAPACITOR)

In some OP AMP and other applications, matched or tracking power supply voltages are required. For the non-critical applications, the high PSRR of the OP AMP can accommodate the slight imbalance in the + and − supply voltages that results from the configuration in Fig. 3.29.

3.7   TRACKING REGULATORS

Assuming that a floating regulated voltage is available, a simple tracking supply can be constructed by adding the supply-splitting circuit shown in Fig. 3.30. The 741 OP AMP is used in the voltage follower mode with $Q1$ and $Q2$

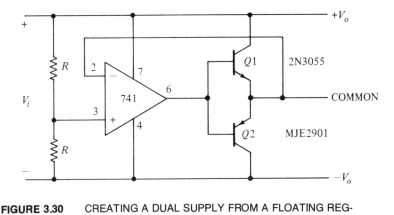

FIGURE 3.30   CREATING A DUAL SUPPLY FROM A FLOATING REG-ULATOR

(power transistors) used to provide the high current for the output. The input to the voltage follower is exactly halfway between the absolute voltages at the two input leads; therefore the two emitters, which are the amplifier output, must also be halfway. The matching of the two resistors determines the matching of the + and − outputs referenced to common. Note that the maximum input voltage is limited by the maximum supply voltage of the 741.

A true tracking regulator is formed by building a conventional positive (or negative) regulator, and then making the other polarity output voltage proportional to the output of the first regulator. One possible scheme using the 558 dual OP AMP is shown in Fig. 3.31. The positive side contains a conventional series

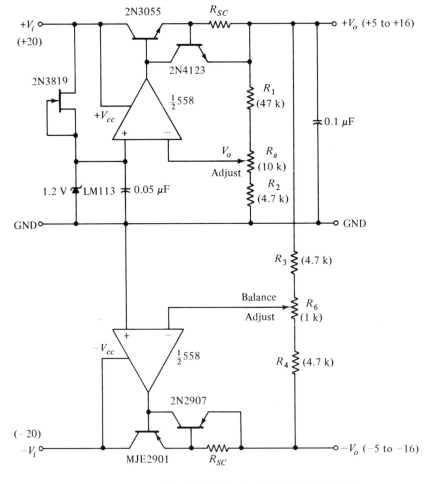

FIGURE 3.31   ADJUSTABLE DUAL TRACKING SUPPLY

regulator with current-limiting and a FET current source feeding a band-gap reference diode. The positive output voltage is adjustable by the R_a potentiometer. For the negative supply, one-half of the 558 is used in the inverting mode with a gain of -1 (adjusted to exactly -1 by the R_b potentiometer). The input voltage is the positive output voltage; thus the negative output must be equal to the negative of the positive output. This is the tracking feature. Should the positive output voltage change for whatever reason, the negative output also changes in direct proportion. Note that if the negative side is shorted to ground and goes into the current-limit mode, the positive output is unaffected. However, the same is not true should the positive side go into the current-limit mode. Since the negative side tracks the positive, as the $+$ side output voltage collapses, so will the $-$ side output voltage. The 558 OP AMP is internally compensated; the bypass capacitor across the reference decreases output noise voltage due to the reference, while the output capacitor reduces output impedance at higher frequencies.

Dual tracking regulators are also available on a single monolithic chip. The output voltages track in the same manner as described above, and are internally preset to ± 12 V, ± 15 V or $+5$ and -12 V. The 325 dual tracking regulator is representative of this type of regulator. While the 325 offers fixed output voltages of $+15$ and -15 V , it is programmable in terms of current-limiting and maximum output current. For electrical characteristics see the Appendix. For a ± 15 V tracking regulator with a current limit of 100 mA, no external components are necessary, as shown in Fig. 3.32. As long as the input voltage is within the ± 18 to ± 30 V limits, the output will be within 10 mV of the nominal voltage over the full line and load regulation range. Obviously, regulators like the 325 are very attractive for OP AMP supplies.

The output current of the 325 can be boosted by the addition of external power transistors, as shown in Fig. 3.33. Under these conditions, the output current is limited externally by the choice of R_{CL}. The current-limit sense voltage for the 325 regulator is shown in Fig. 3.34. As an example, consider a 1 A output current requiring perhaps a 20 to 30 mA output current from the 325. The power dissipation of the 325 with a ± 25 V input is approximately 50 V times the output current or 150 mW. Assuming the N package with a thermal resistance of 150°C/W, we estimate the junction temperature to be:

$$T_J = 25 + (150)(0.15) \cong 48\ °C$$

From Fig. 3.34 at 50°C, we estimate that the $+$ sense voltage is 0.63 V and the $-$ sense voltage 0.52 V. From these values we determine the current-limit resistors:

$$R_{CL+} = \frac{0.63\ V}{1\ A} = 0.63\ \text{Ohms}$$

and

$$R_{CL-} = \frac{0.52\ \text{V}}{1\ \text{A}} = 0.52\ \text{Ohms}$$

Depending on the high frequency characteristics of the particular 2N3055 transistors used, the output capacitance might have to be increased to prevent regulator instability.

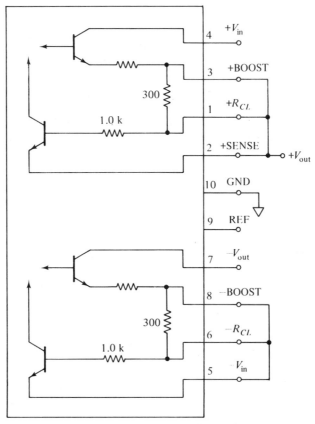

Note: Pin numbers for metal can package only.

FIGURE 3.32   325 ± 15V 100 mA TRACKING REGULATOR

Table 3.3 lists some of the other tracking regulators with their output voltages and rated output currents.

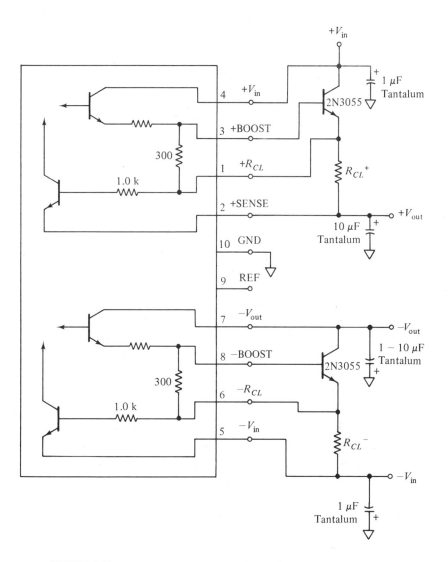

FIGURE 3.33   325 ± 15V HIGH CURRENT TRACKING REGULATOR

3.8   COMPLETE POWER SUPPLY

We now consider the complete dc power supply, typically consisting of an iron-core power transformer, full-wave rectifier, capacitor filter, and voltage regulator.

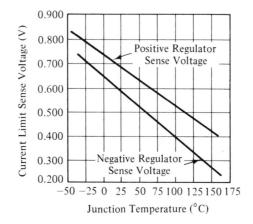

FIGURE 3.34   325 CURRENT-LIMIT SENSE VOLTAGE CHARACTERISTICS

TABLE 3.3   DUAL TRACKING REGULATORS

TYPE NO.	OUTPUT VOLTAGES
LM 325	±15 (fixed)
LM 326	±12 (fixed)
LM 327	+5, −12 (fixed)
MC 1568	±15 (adjustable)
RC 4194	adjustable
RC 4195	±15 (fixed)

Although other configurations using half-wave rectification and more complex filtering are possible, they are not practical. The full-wave rectifier or bridge rectifier is available in a single package of cost comparable to a single diode, so the full-wave rectifier is preferable to a half-wave. Also since the IC voltage regulators offer extremely high ripple rejection, filters more complex than a single capacitor are not necessary.

A full-wave rectifier and filter with a center-tapped secondary transformer, shown in Fig. 3.35, or a bridge rectifier and capacitor filter, shown in Fig. 3.36, are suitable for most single output voltage applications. For dual or tracking regulator applications, a center-tapped transformer and bridge rectifier combination of the type shown in Fig. 3.37 can be used.

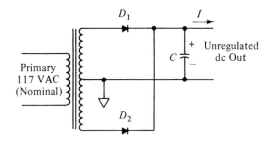

FIGURE 3.35   FULL-WAVE UNREGULATED SUPPLY USING CENTER-TAPPED TRANSFORMER

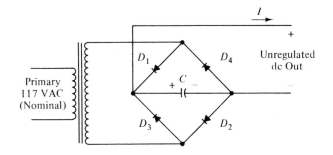

FIGURE 3.36   BRIDGE RECTIFIER UNREGULATED SUPPLY

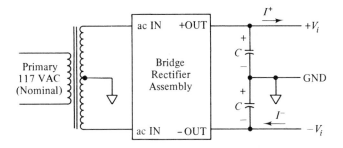

FIGURE 3.37   UNREGULATED SUPPLY FOR DUAL OR TRACKING VOLTAGE REGULATORS

For 60 Hz input, the waveshapes for the rectifier and filter circuits are shown in Fig. 3.38. Note that as soon as any output current is drawn from the filter, the dc voltage decreases and the ripple appears. The higher the output current, the lower the dc (or average) voltage and the higher the ripple. The approximate performance of the circuits is summarized in Table 3.4. (Note that transformer secondary ratings are usually given in V rms, which is $V_M/1.41$.) For the full-

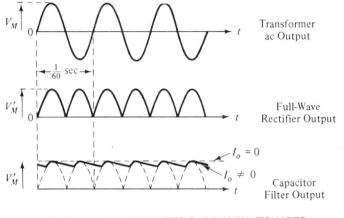

FIGURE 3.38   UNREGULATED SUPPLY WAVESHAPES

TABLE 3.4   SUMMARY OF RECTIFIER CHARACTERISTICS

CIRCUIT FIG. NO.	TRANSFORMER OUTPUT VOLTAGE (peak)	PEAK RECTIFIER OUTPUT VOLTAGE V'_M	FILTER DC OUTPUT VOLTAGE V_{dc}	OUTPUT RIPPLE FACTOR $r = \dfrac{V_{0\ ac\ p\text{-}p}}{V_{dc}}$
3.35	$V_M - 0 - V_M$	$V_M - V_D$	$V'_M - \dfrac{I}{C}(420)^*$	$\dfrac{2400}{C}\dfrac{I^*}{V_{dc}}$
3.36	V_M	$V_M - 2V_D$	$V'_M - \dfrac{I}{C}(420)^*$	$\dfrac{2400}{C}\dfrac{I^*}{V_{dc}}$
3.37	$V_M - 0 - V_M$	$\pm [V_M - V_D]$	$\pm [V'_M - \dfrac{I}{C}(420)]^*$	$\dfrac{2400}{C}\dfrac{I^*}{V_{dc}}$

*For 60Hz input with I in amps, V in volts, and C in microfarads

wave rectifier in Fig. 3.35, the peak output voltage is only one diode drop lower than the peak ac voltage, while in the bridge circuit, the peak output is two diode drops below the peak. It is not untypical to have a 1 V forward diode drop at 1 A, so this voltage must be accounted for in determining the transformer to be used in a particular application.

DETERMINING TRANSFORMER SECONDARY VOLTAGE

From the regulator design we would have the maximum dc current as well as the permissible range of input dc voltages. The first step is to make a tentative choice of filter capacitor. From this value and the nominal dc voltage and maximum dc current we next calculate the approximate ripple voltage. The minimum regulator dc input voltage must equal the dc filter output voltage *minus* one-half the peak-to-peak ripple voltage. This allows us to calculate the absolute minimum secondary voltage for the transformer. In a similar manner, the maximum secondary voltage is determined by making the maximum dc regulator input voltage (at zero current) equal to the maximum peak ac rectifier voltage. The result is a range of secondary voltages—the final choice should be as close to the middle of the range as possible, commensurate with transformer availability. Consider the following example: from the regulator design we have a requirement to maintain the dc input voltage between 7 and 15 Vdc at a current of less than 1 A. We might start by choosing a 1000 μF, 25 V electrolytic capacitor. From Table 3.4 we can solve for the minimum secondary voltage:

$$V_M = V_{i\min} + V_D + 420\,I/C + 1200\,I/C = 9.6 \text{ V}$$

The maximum peak secondary voltage is simply $V_{i\max}$ or 15 V. Thus we need a transformer with a secondary voltage of between 10 V and 15 V peak, or 7 and 11 V rms. Our final choice might be a transformer with a 9 V rms secondary.

DETERMINING THE FILTER CAPACITOR

If a maximum ripple from the regulated supply is specified, we proceed to determine the rectifier ripple by multiplying the output ripple by the regulator ripple-rejection factor. The capacitor value needed is then calculated from Table 3.4. Consider that a 309 regulator is to be used in the example above with a maximum output ripple of 5 mV. From the 309K data sheets we see that the ripple-rejection factor is at least 60 dB (for frequencies below 10 kHz), so this gives a maximum permissible ripple voltage at the 309 input of 5 V. With a maximum of 1 A and a nominal input voltage of 10 V, we calculate from Table 3.4 the minimum capacitor needed as 2400 $I/5V$ = 480 μF.

DETERMINING LOAD REGULATION

The load regulation for the regulator is usually specified. For the 309 it is listed as a typically 50 mV change in output voltage for a change in load current of 1.5 A. However, since the filter dc voltage drops at higher currents, we need to account for the line regulation of the regulator in determining the load regulation of the supply. The line regulation for the 309 is listed as 50 mV maximum change in output voltage for an input change from 7 to 25 V, or approximately 3 mV/V. If a 1000 μF filter capacitor is used and the maximum load current is 1 A, the filter dc voltage change due to a load current change of 1 A is $I420/C$ = 0.42 V. This causes the output voltage to vary by approximately 1 mV. Thus the power supply load regulation is a maximum of 51 mV for a load current change of 1 A.

HEAT SINKS

In some applications, the IC regulator can be used in the dc supply without a heat sink, typically where the current requirements are low or when each of a number of regulators supplies only a small part of the whole system.

However, more reliable operation resulting from lower junction operating temperatures can be achieved with the use of an appropriate heat sink. The choice of the heat sink is strictly dictated by the power to be dissipated in the regulator and/or external power transistor and the desired maximum operating junction temperature.

For devices dissipating more than a few watts, the heat sink should be mounted on the outside of the metal enclosure. For lower dissipation, the heat sink can be mounted directly on the printed circuit board. Typical examples of heat sinks and their uses are illustrated in Figs. 3.39, 3.40, and 3.41. In Fig. 3.39 an example of a discrete component regulator is shown. Note the three types of heat sinks used: in the upper right, a vertical heat sink for plastic power packages; in the middle, a heat sink for a small-signal plastic transistor; and in the foreground, a copper heat sink formed by the foil pattern on the pc board itself. Figure 3.40 shows different sizes of heat sinks that accommodate either the K or T power package and are suitable for either chassis or pc board installations. Figure 3.41 shows the implementation of $+5$, $+15$, and -15 V three-terminal regulators.

FIGURE 3.39   TYPICAL DISCRETE COMPONENT REGULATOR

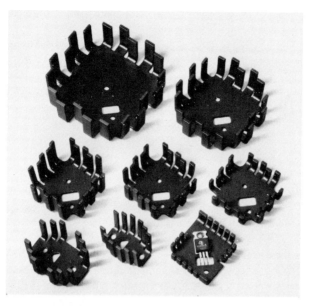

FIGURE 3.40   HEAT SINKS FOR IC VOLTAGE REGULATORS

FIGURE 3.41   EXAMPLE OF ON-BOARD REGULATORS WITH HEAT SINKS

Specific heat sink characteristics are available from heat sink manufacturers. We shall use the IERC (International Electronic Research Corporation, Burbank, California) type UP2−T03−B heat sink mated to a 309K (TO−3 package) as an example. The heat sink is listed as having a thermal resistance of approximately 7°C/W (see Appendix). The junction-to-case thermal resistance of the 309K is listed as 3°C/W. Let us assume an input voltage of 10 V (output at 5 V), and an output current of 1 A. The power dissipation at maximum load is 5 W. The thermal resistance junction-to-ambient Θ_{JA} is:

$$\Theta_{JA} = \Theta_{JC} + \Theta_{CA} = 3 + 7 = 10°C/W$$

The rise in junction temperature above ambient due to the dissipation of 5 W is then $P\Theta_{JA} = 50°C$, giving an operating junction temperature of approximately 75°C, which is well below the maximum of 125°C. (Ambient temperature is assumed to be 25°C.)

Without a heat sink, the junction-to-ambient thermal resistance of the 309 K is 35°C/W. At the maximum junction temperature of 125°C without a heat sink, the maximum power dissipation is determined to be approximately 2.85 W. However, the maximum device dissipation listed by the manufacturer is only 2 W. Thus with no heat sink and a 10 V input, the output current of the 309K is limited to a maximum of 400 mA.

In practice, the minimum thermal resistance of the heat sink is determined from the desired power dissipation for the regulator and its junction-to-case thermal resistance, using the maximum junction temperature for the regulator. The appropriate heat sink is then chosen on the basis of the power it has to dissipate and the maximum thermal resistance required. It is usually a good idea to be conservative in thermal designs, that is, to operate all devices well below their thermal limits.

In the natural convection (as opposed to forced air) application of heat sinks, they should be mounted so that the largest of the convective surfaces is in the vertical plane. In addition, if the heat sink is inside the enclosure, venting of the enclosure will aid in the efficiency of the heat sink.

POWER SUPPLY SPECIFICATIONS   Lastly, we consider determining the power supply specifications for a given system's implementation. If the system's voltage and current requirements are not known ahead of time, choose readily available voltages, trying to keep the number of different supply voltages to a minimum. Decide on one of the two methods: use either a single central power supply to power all of the circuits or a single unregulated supply with individual voltage regulators powering different sections of the system. (Unless there are overriding considerations, the second approach is preferable.) If a central supply is to be used, the system without a power supply should be implemented first. A bench supply can be used for testing and troubleshooting. Once the system is finalized, the current drain on the bench supply should be measured to determine the system's current demand. The power supply for the system can now be designed—it is usually a good idea to provide approximately 20% excess current capability.

If individual on-card regulators are to be used, the system should be segmented into functional units. Then, using manufacturers' data sheets, the current demand for each subsystem can be estimated to determine the current required of the regulator. In on-card regulator applications, it is advisable to use two or three low current regulators rather than one high current. In addition, the regulators should (if practical) be spaced as far apart as possible. In this manner, the heat generated will be more uniformly distributed, resulting in a lower overall circuit temperature.

A commonly overlooked consideration in power supply design is the physical connection made between the transformer, the rectifier and filter, and the regulator and the circuit to be powered. Keeping in mind the current level, use wire and printed circuit traces of appropriate size. For 2 oz. copper printed circuit boards the trace resistance per linear inch is 0.000227/W (in Ohms). For example, a trace 4″ long and 0.050″ wide has a resistance of approximately 18 milliOhms. In most cases this is sufficiently small. However, consider this trace to be the supply line carrying 3 A. Then the voltage drop resulting would be in excess of 50 mV in each line (assuming equal lengths for the supply and return). Note that this significantly degrades the regulation. The trace impedance also increases at high frequencies, so in general, the power supply line widths on printed circuit boards should be made as wide as practical. This is especially true for the ground or common trace—unwanted ground loops are thus minimized.

REVIEW QUESTIONS

3.1 A voltage regulator is listed as having a line regulation of 0.5% or better. Explain this specification and specify the expected variation in output voltage per 1 V change in the input voltage.

3.2 The circuit to be supplied by a voltage regulator must have a ripple of no more than 50 mV peak-to-peak. The raw dc has 7 V p-p ripple. What must be the regulator ripple-rejection ratio?

3.3 What is the significance of the dropout voltage? What is it for the 309 regulator?

3.4 What is the quiescent (standby) current for a 309 regulator?

3.5 A type 723 regulator is to provide a 12 V output at 50 mA. The input is normally 17 V; over line and load variations, the input is between 15 and 20 V with a ripple of 1 V maximum. Specify the proper circuit values (making certain that the maximum power dissipation of the regulator is not exceeded under *any* load conditions); also specify the line regulation and output ripple.

3.6   What are the advantages of fold-back current-limiting over conventional current-limiting?

3.7   What are the considerations in choosing a fixed three-terminal regulator over an adjustable one (like the 723)? When would an adjustable regulator be preferable?

3.8   What is the minimum input voltage for a 309 regulator? If the input voltage is 10 V dc minimum, what is the peak ripple that can be tolerated for the 309 regulator to function properly?

3.9   Determine the filter capacitor to be used with a 309 regulator if the output ripple is to be 100 mV maximum with a 1 A output current.

3.10   An LM309K is used in a power supply with a maximum output current of 750 mA. The average input voltage is 12 V. Determine if a heat sink is needed, and, if so, what minimum thermal resistance it should have.

3.11   Repeat the above example if the output current reaches peaks of 1.2 A for a time no longer than 20 ms, while it is normally at 0.5 A for at least 120 ms.

3.12   Make a comparison of central power supplies and individual on-card regulators. (List the advantages and disadvantages of each.)

3.13   What are the most critical physical layout properties for a power supply?

4

TTL

Transistor-Transistor-Logic (TTL or T²L) integrated circuits are available in five families and in complexity ranging from memories to simple inverted circuits. They are designed to be used as digital building-blocks, thus allowing for easy implementation of any digital function, be it complex or simple.

Most IC manufacturers offer TTL circuits; therefore they are readily available off-the-shelf from just about all distributors. In addition, the input and output characteristics are standardized among all the manufacturers as is the numbering system (with very few exceptions), making possible total interchangeability and ease of procurement. As is the case with all ICs, the cost of TTL ICs has been reduced over the years, so that now, for example, a *NAND* gate costs less than a socket that might be used, but need not be used, to hold it. The low cost plus the versatility offered by the full line of logic functions makes the TTL family an extremely attractive one in most general digital applications.

4.1 TTL FAMILIES AND CHARACTERISTICS

The standardized numbering system used for TTL ICs is the following: depending on the manufacturer, typically two letters followed by either 54 or 74, followed by an additional letter and/or numbers. The first two letters are generic to the manufacturer only and carry no information as to the function. The next two numbers signify TTL: if they are 74, they signify commercial TTL (temperature range 0 to 70°C); if they are 54, they signify military temperature range TTL (−55 to 125°C). In the next position, there may or may not be a letter. If there is none, general TTL family is specified. If the letter is L, the low-power TTL

115

family is specified; similarly, H stands for high-speed TTL, S stands for Schottky TTL, and finally, LS stands for low-power Schottky TTL. The next two or three digits actually specify the logic function for that particular IC. Lastly, a letter suffix denotes the type of package. For example, an IC labeled SN74LS20N is a dual four-input *NAND* gate (signified by 20), commercial temperature range TTL (signified by 74), made by Texas Instruments Inc. (signified by SN), in a plastic 14 pin dual-in-line package (DIP) (signified by the N suffix), and it is low-power Schottky TTL gate.

Each of the subfamilies within the TTL family serves to fill a specific characteristic making it more suitable for a given application. When there are no special requirements, the general or standard TTL series should be used. When low power consumption is required and lower operating speed can be tolerated, a low-power TTL should be used. Where high switching speeds are required and power consumption is not critical, a high-speed (H) or Schottky TTL should be used. Where low power consumption as well as fast switching speed is required, the low-power Schottky TTL should be used. The specific switching speed as well as power consumption for the different subfamilies is listed in Table 4.1. (In applications requiring extremely low power consumption and where lower switching speeds are to be tolerated, the CMOS logic family discussed in the next section should be considered.)

TABLE 4.1   COMPARISON OF TTL FAMILIES

SERIES	TYPICAL TOTAL PROPAGATION DELAY PER GATE	TYPICAL POWER DISSIPATION PER GATE
54/74	18 ns	10 mW
54H/74H	12 ns	23 mW
54L/74L	66 ns	1 mW
54S/74S	6 ns	19 mW
54LS/74LS	19 ns	2 mW

It is quite possible to use more than one of the subfamilies of TTL in a single system. An example of such a case is where a high frequency signal is to be divided down to a lower frequency. Here we might use a Schottky TTL in the front end of the divider chain, followed by a standard TTL once the frequency has been scaled down. However, when mixing different families, it is essential to observe the maximum loading characteristics of the particular circuits involved.

To more fully understand and appreciate the characteristics of the different TTL families, let us consider the circuits inside the IC. For simplicity, let us consider the quad two-input *NAND* gate circuits; each 7400 IC contains four (quad) separate *NAND* gates. The circuit diagram for one of these gates is shown in Fig. 4.1. A number of circuit features shown here are characteristic of the

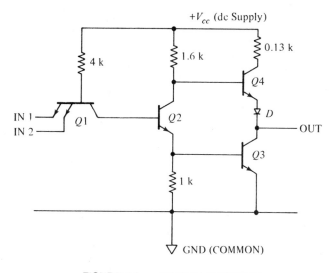

FIGURE 4.1   54/7400 SCHEMATIC

entire standard TTL family: first, the multiple-emitter transistors with a 4 k resistor to the supply; next, the output configuration of $Q3$, D, and $Q4$, called the totem-pole output.

To see the basic operation, let us connect the two inputs together (thus forming an inverter) and apply a low voltage to the input. Under these conditions, the base-emitter junction of $Q1$ is on and conducts, sourcing current from the input. This brings the base of $Q1$ to a voltage low enough that the diodes formed by the collector-base junction of $Q1$ and the base-emitter junctions of $Q2$ and $Q4$ are not sufficiently forward biased to be able to conduct. The situation is shown in Fig. 4.2. The base of $Q3$ is near the 5 V supply, and therefore $Q3$ and D are forward biased and are conditionally on. Without a load to ground from the output, the current through $Q3$ and D is quite small. Thus we would expect an output voltage of approximately $5-2(0.7) = 3.6$ V. Note that for a low input we have a high output, the inverter function. If we assume that the three junctions of $Q1$, $Q2$, and $Q3$ can be forward biased by as much as 0.5 V and still not conduct significantly, we see that the highest voltage at the base of $Q1$ is 1.5 V. This forces us to apply an input of $1.5 - 0.7 = 0.8$ V or lower to maintain the output in the high state. Thus, the maximum low-level input voltage, V_{IL}, should be about 0.8 V. If we assume the input low voltage to be typically 0.3 V, we can calculate the input current I_{IL}, when a low is applied as $(5 - 0.7 - 0.3)/4 = 1$ mA.

When a high level is applied to the input, the base-emitter junction of $Q1$ is off and $Q2$ and $Q3$ are on. This situation is shown in Fig. 4.3. Assuming the voltage drop across the saturated junctions of $Q1$, $Q2$, and $Q3$ to be 0.8 V, we see that the base of $Q1$ is at $3(0.8V) = 2.4$ V. The drop across the 4 k resistor is

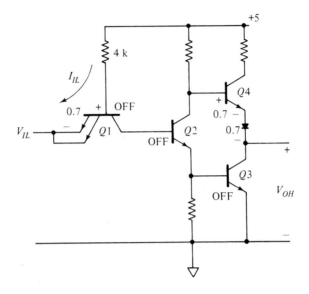

FIGURE 4.2   TTL GATE WITH OUTPUT HIGH

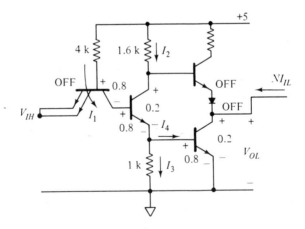

FIGURE 4.3   TTL GATE WITH OUTPUT LOW

then $5 - 2.4 = 2.6$ V, giving the current $I_1 = 0.65$ mA. With $Q3$ saturated, its base is at 0.8 V; therefore if the collector-emitter saturation voltage of $Q2$ is 0.2 V, the collector of $Q2$ is 1 V above ground. We can now see the purpose of the diode, for if the diode were not in the circuit, the base-emitter voltage of $Q4$ would now be $1 - 0.2 = 0.8$ V and $Q4$ would be biased on. Since the net voltage across the diode and the base-emitter junction of $Q4$ is 0.8 V, both are effectively off. The current I_2 is:

$$I_2 = \frac{5 - 1}{1.6} \text{ mA} \cong 2.5 \text{ mA}$$

Thus, in order to insure saturation of $Q2$, since I_2 is the collector current of $Q2$, the minimum β is $I_{C_2}/I_{B_2} = 2.5/0.65 = 3.8$. Typically β is larger than 20; therefore the assumption of $Q2$ being saturated is justified. The emitter current of $Q2$ is the sum of its base and collector currents: $0.65 + 2.5 = 3.15$ mA. Since the drop across the 1 k resistor is 0.8 V, the current I_3 is 0.8 mA. The base current of $Q3$, I_4, is the difference between the emitter current of $Q2$ and I_3: $3.15 - 0.8 = 2.35$ mA. With no load at the output, the only current that flows into the collector of $Q3$ is the very small leakage current from $Q4$ and D. Therefore, $Q3$ is definitely saturated. However, consider that the output in the low state has N inputs of similar gates connected to it. As we saw previously, when a low is applied to an input, it sources approximately 1 mA. Therefore, $Q3$ must be capable of sinking this current. For N gates at the output, the net current into the collector of $Q3$ is NI_{IL} or N mA. If, for example, $N = 10$, then the collector current of $Q3$ would have to be 10 mA. This would require a minimum β for $Q3$ of $10/2.35 = 4.3$. The number N has a value of 10 for the standard TTL family and is the *fan-out*—the number of similar gates that can be connected to the output while still maintaining the proper logic level. For TTL gates, the fan-out limitation is dictated when the output is in a low state and has to sink the input current of the loading gates. When the output is in a high state, the gates connected to it offer a high impedance due to the fact that their base-emitter junctions are essentially off.

We can get an idea of the minimum input voltage that is recognized as a high by noting that the base of $Q1$ is three diode drops above ground or approximately 2.4 V. The base-emitter junction of $Q1$ must be essentially off; therefore it cannot be more forward biased than 0.5 V. This gives:

$$V_{IH \text{ min}} = 2.4 - 0.5 = 1.9 \text{ V}$$

Under static operation, i.e. with an input voltage fixed at either a low or a high, the input transistor functions as three diodes. To understand why a transistor is used here, consider the sequence of events when the input is brought low (having previously been high). It takes a non-zero time to remove the charge of the forward biased base-emitter junctions of $Q2$ and $Q3$. If we approximate these two junctions by a capacitor while the transistors are turning off, the equivalent circuit shown in Fig. 4.4 results. If the input is at 0.2 V, the base of $Q1$ is at 0.9 V; thus the collector-base junction of $Q1$ is reverse biased. The transistor is now in its linear active region where $I_C = \beta I_B$. Since the collector current discharges the effective capacitance, there is an improvement in switching speed over the circuit using diodes instead of $Q1$. (Diode-transistor-logic, DTL, circuits are constructed with $Q1$ replaced by individual diodes.) This improvement is due to

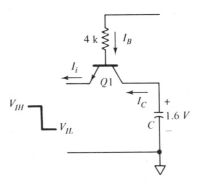

FIGURE 4.4   INPUT OF TTL GATE DURING HIGH-TO-LOW TRANSITION

the current gain β of the transistor increasing the discharge current. Note that during this discharge time, which may be only a few nanoseconds, the current sourced by the input is significantly higher than the value calculated under static conditions.

As the input changes from a high to a low, transistors $Q2$ and $Q3$ turn off, while $Q4$ and D turn on. However, $Q4$ and D begin to conduct slightly before $Q3$ is fully off. This provides a current path through $Q4$, D, and $Q3$ to ground, causing a momentary but significant current pulse to appear on the power supply line. This and other switching phenomena cause high-frequency "spikes" on the supply line which require capacitive bypassing. Typically, the bypass capacitor is between 0.01 and 0.1 μF. To see the need for the combination of $Q4$ and D, as opposed to simply a resistor from the collector of $Q3$ to +5 V, consider the loading of the output by other gates. If the output is in the low state, these gates are all sinking current into $Q3$. When $Q3$ turns off, there is an effective capacitance seen at the output due to the charge stored in the base-emitter junctions of the input transistors of succeeding gates. This situation is represented by the circuit shown in Fig. 4.5. Under these conditions, $Q4$ saturates momentarily, providing a high output current to "pull up" the output voltage. The current with $Q4$ saturated may be estimated by assuming the capacitor to be at 0.2 V, the diode on voltage as 0.7 V, and $V_{CE4} = 0.2$ V. Thus $I_E = (5 - 0.2 - 0.7 - 0.2)/0.13$ mA = 30 mA. This current decreases to essentially zero very rapidly as the effective capacitance charges up and the output voltage attains its static value. The reason for calling this output configuration an active pull-up should be obvious.

We can compare our calculations with the typical parameters listed by the manufacturer which are shown in Fig. 4.6. The listed values for V_{IH} and V_{IL} of 2 V and 0.8 V agree closely with our calculated values of 1.9 V and 0.8 V. Also, the maximum value for I_{IL} of 1.6 mA compares favorably with our calculated typical value of 1 mA. Note also that the basic parameters are relatively independent of the specific gate, but are a function of the family type.

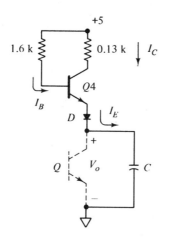

FIGURE 4.5   OUTPUT OF TTL GATE DURING LOW-TO-HIGH TRANSITION

The circuit for a low-power version of the two-input *NAND* gate is shown in Fig. 4.7. Note that the circuit is identical for the 54/7400 and the 54L/74L00, with the only difference being the scaling of resistors. Since each of the resistors in the low-power gate is roughly one-tenth the value of resistors of the standard gate, we expect all the static currents in the 'L00 gate to be one-tenth as well. Thus the power dissipation for the 'L00 gate is lower (but not by a factor of 10) than that of the '00 gate. While power dissipation is decreased (a desirable effect), the lower current levels mean that any effective capacitance (like that of transistor junctions) takes a longer time to charge and discharge. Thus the switching speed is degraded.

The fan-out within the low-power family, i.e. driving other low-power gates, is 20. However, when a low-power gate is used to drive a standard gate, the fan-out is only 2 due to the higher input current sourced by the standard gate. Should the fan-out be exceeded, the output transistor may not be saturated and an unallowed output voltage may result. This condition must at all times be avoided. The fan-out for standard TTL driving low-power is typically 40, that is, a standard TTL gate can drive 40 low-power gates connected to its output without degradation of the output voltage levels.

The circuit schematic for a high-speed gate, 54H/74H00, is shown in Fig. 4.8. Note that when compared to the '00 gate, the major difference is the Darlington-connected transistors $Q4$ and $Q5$. Resistor values are all scaled down, thus increasing the quiescent currents as well as the power dissipation over those in the standard gates. The reverse of the conditions that apply to the low-power gate is in effect here: since the current levels are higher, switching speed is improved at the cost of higher power dissipation. Static operation of the 'H00 gate is exactly the same as the '00 gate, with the base-emitter junction of $Q5$

recommended operating conditions

PARAMETER		SERIES 54 / SERIES 74 ('00, '04, '10, '20, '30) MIN	NOM	MAX	SERIES 54H / 74H ('H00, 'H04, 'H10, 'H20, 'H30) MIN	NOM	MAX	SERIES 54L / 74L ('L00, 'L04, 'L10, 'L20, 'L30) MIN	NOM	MAX	SERIES 54LS / 74LS ('LS00, 'LS04, 'LS10, 'LS20, 'LS30) MIN	NOM	MAX	SERIES 54S / 74S ('S00, 'S04, 'S10, 'S20, 'S133) MIN	NOM	MAX	UNIT
Supply voltage, VCC	54 Family	4.5	5	5.5	4.5	5	5.5	4.5	5	5.5	4.5	5	5.5	4.5	5	5.5	V
	74 Family	4.75	5	5.25	4.75	5	5.25	4.75	5	5.25	4.75	5	5.25	4.75	5	5.25	
High-level output current, IOH	54 Family			−400			−500			−100			−400			−1000	µA
	74 Family			−400			−500			−200			−400			−1000	
Low-level output current, IOL	54 Family			16			20			2			4			20	mA
	74 Family			16			20			3.6			8			20	
Operating free-air temperature, TA	54 Family	−55		125	−55		125	−55		125	−55		125	−55		125	°C
	74 Family	0		70	0		70	0		70	0		70	0		70	

electrical characteristics over recommended operating free-air temperature range (unless otherwise noted)

PARAMETER	TEST FIGURE	TEST CONDITIONS†		SERIES 54 / 74 ('00, '04, '10, '20, '30) MIN	TYP‡	MAX	SERIES 54H / 74H ('H00, 'H04, 'H10, 'H20, 'H30) MIN	TYP‡	MAX	SERIES 54L / 74L ('L00, 'L04, 'L10, 'L20, 'L30) MIN	TYP‡	MAX	SERIES 54LS / 74LS ('LS00, 'LS04, 'LS10, 'LS20, 'LS30) MIN	TYP‡	MAX	SERIES 54S / 74S ('S00, 'S04, 'S10, 'S20, 'S133) MIN	TYP‡	MAX	UNIT
VIH High-level input voltage	1, 2			2			2			2			2			2			V
VIL Low-level input voltage	1, 2	54 Family				0.8			0.8			0.7			0.7			0.8	V
		74 Family				0.8			0.8			0.7			0.8			0.8	
VI Input clamp voltage	3	VCC = MIN, II = §				−1.5			−1.5		*	*			−1.5			−1.2	V
VOH High-level output voltage	1	VCC = MIN, VIL = VIL max, IOH = MIN	54 Family	2.4	3.4		2.4	3.5		2.4	3.3		2.5	3.4		2.5	3.4		V
			74 Family	2.4	3.4		2.4	3.5		2.4	3.2		2.7	3.4		2.7	3.4		
VOL Low-level output voltage	2	VCC = MIN, VIH = 2 V, IOL = MAX	54 Family		0.2	0.4		0.2	0.4		0.15	0.3		0.25	0.4			0.5	V
			74 Family		0.2	0.4		0.2	0.4		0.2	0.4		0.35	0.5			0.5	
II Input current at maximum input voltage	4	VCC = MAX, VI = 5.5 V				1			1			0.1			0.1			1	mA
IIH High-level input current	4	VCC = MAX	VIH = 2.4 V			40			50			10			20			50	µA
			VIH = 2.7 V																
IIL Low-level input current	5	VCC = MAX	VIL = 0.3 V ('LS30)			−1.6			−2			−0.18			−0.4			−2	mA
			VIL = 0.4 V (Others)												−0.36				
			VIL = 0.4 V																
			VIL = 0.5 V																
IOS Short-circuit output current¶	6	VCC = MAX	54 Family	−20		−55	−40		−100	−3		−15	−40			−40		−100	mA
			74 Family	−18		−55	−40		−100	−3		−15	−42			−40		−100	
ICC Supply current	7	VCC = MAX									See table on next page									mA

†For conditions shown as MIN or MAX, use the appropriate value specified under recommended operating conditions.
‡All typical values are at VCC = 5 V, TA = 25°C.
§ II = −12 mA for SN54'/SN74', −8 mA for SN54H'/SN74H', and −18 mA for SN54LS'/SN74LS' and SN54S'/SN74S'.
¶Not more than one output should be shorted at a time, and for SN54H'/SN74H' and SN54S'/SN74S', duration of short-circuit should not exceed 1 second.
*The input clamp voltage specification is effective for Series 54/74 and 54H/74H parts date-coded 7332 or higher.

TEXAS INSTRUMENTS
INCORPORATED
POST OFFICE BOX 5012 • DALLAS, TEXAS 75222

FIGURE 4.6   TTL DATA SHEET

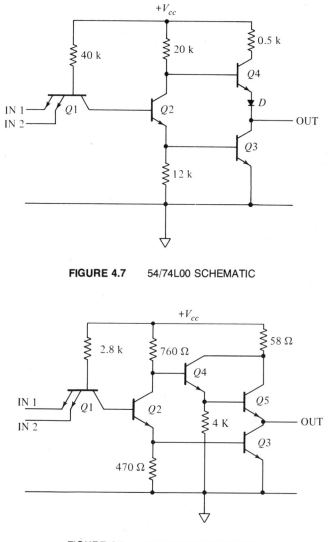

FIGURE 4.7   54/74L00 SCHEMATIC

FIGURE 4.8   54/74H00 SCHEMATIC

replacing the diode in the '00 gate. While the output is undergoing a transition from a low to a high, the Darlington configuration of $Q4$ and $Q5$ turns on and provides current to pull up any capacitive load. The circuit during this transition is shown in Fig. 4.9. The charging current is I_{E5}; and with a lower collector resistor, the initial charging current is higher than in standard TTL. As the output voltage starts coming up, with both $Q4$ and $Q5$ active, the charging current is approximately β $(\beta I_{B5} - I)$. This is higher than the standard TTL current by a factor almost equal to the β of the additional transistor used.

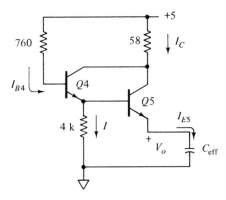

FIGURE 4.9   HIGH SPEED TTL GATE OUTPUT DURING LOW-TO-HIGH
TRANSITION

As is the case with most TTL families, the fan-out for the high-speed gates driving other high-speed gates is 10. The fan-out from high-speed to standard is 12, and from standard TTL to high-speed, 8.

The circuit diagram for a 54S/74S00 gate is shown in Fig. 4.10. Note the transistor symbol: it indicates a Schottky-diode clamp placed across the collector-base junction of a regular *NPN* transistor. The Schottky diode is fabri-

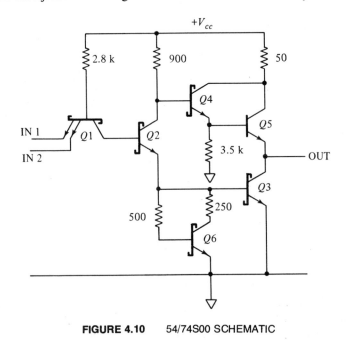

FIGURE 4.10   54/74S00 SCHEMATIC

cated by depositing a metalization layer directly over the collector region, and tying it back to the base. The contact of the metal and the collector region forms a rectifying junction, with a forward drop of 0.4 V. The effect of the clamp across the collector-base junction is to keep the transistor from saturating when it is biased to be on. Consider the conventional transistor in saturation, as shown in Fig. 4.11(a). Depending on how deeply it is driven into saturation, the collector-base voltage is 0.7 V or just slightly higher. For the Schottky-clamped transistor, shown in Fig. 4.11(b), the clamping diode conducts at 0.4 V forward bias, thus not allowing the collector-base voltage to reach its saturated value. As we discussed previously, the more saturated the transistor, the longer it takes to turn off and the slower the switching speed. Thus clamping the transistors and making certain that they do not saturate when on will significantly improve switching speed. Note that other than the use of Schottky-clamped transistors, the circuit configuration of the 'S00 gate is basically the same as the other TTL families already discussed. The only parameter which is different is the V_{OL}, which is typically 0.4 V rather than 0.1 V. However, the Schottky TTL family is still fully compatible with all other TTL families since its maximum V_{OL} is still less than the specified maximum V_{IL}. The resistor values used in the 'S00 gate are very similar to those of the 'H00 gate, so power dissipation is also similar.

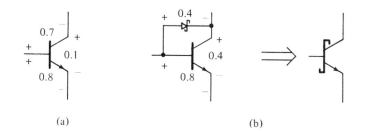

(a)   (b)

FIGURE 4.11   (a) CONVENTIONAL TRANSISTOR IN SATURATION; (b) SCHOTTKY-CLAMPED TRANSISTOR IN THE *ON* STATE

Last, but certainly not least, we consider the low-power Schottky family, whose circuit we see in Fig. 4.12. The circuit is somewhat different from other TTL families while still retaining the basic TTL features: when a low is applied to the input, it sources current through the input diodes; the regular TTL totem-pole or active pull-up comprised of $Q3$ and the Darlington connected $Q4$ and $Q5$ is evident at the output. Note, however, that there are only two junctions (base-emitter diodes of $Q1$ and $Q3$) between the input diode and ground—thus somewhat lower $V_{IH\ min}$ and $V_{IL\ max}$ values might be expected. The 54LS/74LS series is still fully compatible with the other TTL families.

Although we have considered the low-power Schottky TTL family last, it seems to offer the best of all worlds. Note that the resistor values are significantly

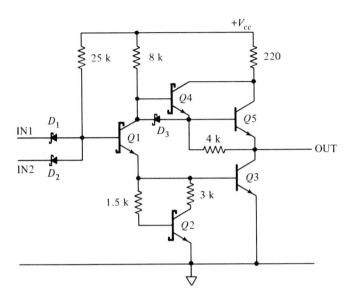

FIGURE 4.12   54/74LS00 SCHEMATIC

higher than their counterparts in the Schottky series, thus providing significantly
lower power dissipation. At the same time, the use of Schottky-clamped transistors improves the switching speed. The low-power Schottky series is functionally
equivalent to standard TTL, even in switching speed, while at the same time
consuming typically one-fifth as much power as standard TTL. These features
make the low-power Schottky family extremely attractive and promise to make it
the TTL family most widely used in the very near future.

In summary, the key things to remember are these. Observe the fan-out
limitation (for convenience, a summary is given in Table 4.2), especially when

TABLE 4.2   FAN-OUT FOR TTL FAMILIES

FROM \ TO	STANDARD 54/74	LOW-POWER 54L/74L	HIGH-SPEED 54H/74H	SCHOTTKY 54S/74S	LOW-POWER SCHOTTKY 54LS/74LS
STANDARD 54/74	10	40	8	8	20
LOW POWER 54L/74L	2	20	1	1	10
HIGH-SPEED 54H/74H	12	50	10	10	25
SCHOTTKY 54S/74S	12	100	10	10	50
LOW-POWER SCHOTTKY 54LS/74LS	5	40	4	4	20

driving loads other than other gates. When interfacing to TTL, remember that, for a low input, all TTL gates require the sinking of current, I_{IL}, to ground. Thus the maximum resistance from an input to ground is $V_{IL\ max}/I_{IL}$. Most manufacturers include a diode from each input to ground. This diode does not in any way affect the operation of the gate; it is included to prevent the input from going more negative than 0.7 V (when the diode conducts). To determine the maximum safe output sink current (output low), determine I_{IL} and multiply it by the fan-out. The maximum output source current (output high) is determined by considering the active pull-up as discussed above and by making certain that the maximum power dissipation of the gate, as well as of the whole IC, is not exceeded. Note that drawing current from a high output will lower the output voltage.

4.2 GATES

The basic logic symbols for inverters and *NAND* and *NOR* gates are shown in Fig. 4.13. The circle on the output of the *NAND* and *NOR* gates is an inversion symbol; thus an *AND* gate symbol is the same as the *NAND* without the little circle on the output. The symbol for an *OR* gate is similarly obtained by leaving out the circle at the output of the *NOR* gate.

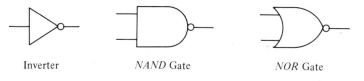

Inverter   *NAND* Gate   *NOR* Gate

FIGURE 4.13   LOGIC GATE SYMBOLS

We shall briefly review the basic logic functions. In an *OR* gate, the output is low only if all inputs are low; otherwise a high output results. Since an inverter gives the opposite logic level at the output, i.e., a high out for a low in and vice versa, a *NOR* gate gives a high output only when all its inputs are low; otherwise, a low output results. An *AND* gate must have all its inputs high to give a high output; otherwise the output is low. Thus, a *NAND* gate gives a low output only if all its inputs are high; otherwise a high output results. This operation is summarized in Table 4.3.

A perusal of any manufacturer's data book will verify that TTL gates of just about any type and fan-in (number of inputs) are available. In most cases, two or more of the same type of gate are available on the same chip. For example, four gates of the type discussed in the previous section are packaged in each 7400 IC. Due to the circuit used to implement gates, as well as the fact that with *NAND*

TABLE 4.3   LOGIC TRUTH TABLES

IN A	IN B	OR GATE	OUTPUT NOR GATE	AND GATE	NAND GATE
LO	LO	LO	HI	LO	HI
LO	HI	HI	LO	LO	HI
HI	LO	HI	LO	LO	HI
HI	HI	HI	LO	HI	LO

gates alone any given logic function can be implemented, *NAND* gates are available in a larger variety than any other type of gate.

In implementing any given logic function, the optimum design does not necessarily utilize the minimum number of gates; the optimum design, one that is lowest in cost, utilizes the minimum number of chips (ICs). To see this, consider the case where one two-input *NAND* gate is necessary. We would use a '00 gate—but note that we get four such gates in the package. Thus if we also needed an inverter, rather than using one-sixth of a type '04 hex inverter, we should implement the inverter with one of the unused *NAND* gates in the '00 IC. This may be done by connecting the two inputs together, as shown in Fig. 4.14(a), or by connecting one input to the +5 V supply and using the other input, as shown in Fig. 4.14(b). The two methods are logically equivalent since both provide the inverter function. However, where switching speed needs to be maximized, the circuit of Fig. 4.14(b) is preferable due to the slightly lower effective capacitance at the input.

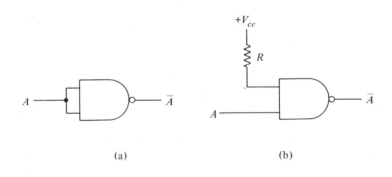

(a)   (b)

FIGURE 4.14   INVERTER USING *NAND* LOGIC

An *AND* gate can be implemented by inverting the output of a *NAND* gate as shown in Fig. 4.15. Instead of an actual inverter, another *NAND* gate may be used.

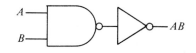

FIGURE 4.15   *AND* GATE USING *NAND* LOGIC

Implementation of an *OR* gate using a *NAND* gate and two inverters is shown in Fig. 4.16. As before, the inverters could be implemented by *NAND* gates themselves. Also note that the *OR* function could be obtained by inverting the output of a *NOR* gate. The inputs of a *NOR* gate could also be connected together to implement an inverter; alternately, one input could be connected to ground, with the other used for the inverter function.

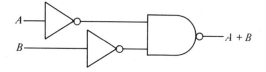

FIGURE 4.16   *OR* GATE USING *NAND* LOGIC

It should be evident that there are almost infinite possibilities for implementing a logic function. However, a little thought coupled with some ingenuity will usually pay off in a logic implementation using the lowest number of ICs as well as being the lowest in cost.

Probably the hardest part of using TTL circuits is properly interfacing other parts of the circuit to the logic part. We shall take up the more general interfacing problem in a subsequent section; now we shall consider the problem of interfacing an input to TTL logic circuits. A number of versions of a TTL *NAND*-type Schmitt trigger are available. One example, together with the logic symbol, is shown in Fig. 4.17. Note the typical TTL input and output structure—thus the circuit readily interfaces with other TTL gates. The major difference in the operation is due to the positive feedback between $Q1$ and $Q2$ provided through R_E. This positive feedback provides hysteresis in the transfer characteristics (see Section 2.11). Typical operation is to buffer a slow and noisy input with a *NAND*-type Schmitt trigger before applying it to any TTL inputs. The characteristics of *NAND*-type Schmitt triggers typically give a low output when the input goes more positive than 1.7 V (V_{T+}) and give a high output when the input falls below 0.9 V (V_{T-}), providing 0.8 V hysteresis. Figure 4.18 shows a typical noisy and slowly varying analog input together with the resulting Schmitt trigger output.

A Schmitt trigger can be easily converted into a square-wave oscillator (see Section 2.12). A simple oscillator configuration is shown in Fig. 4.19. When the

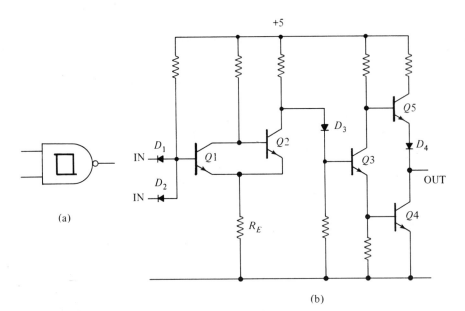

FIGURE 4.17   TTL *NAND* SCHMITT TRIGGER: (a) SYMBOL; (b) TYPICAL
SCHEMATIC (132)

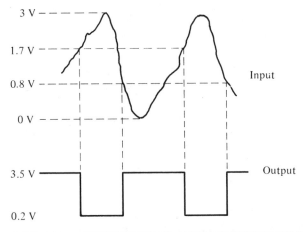

FIGURE 4.18   EXAMPLE OF TTL *NAND* SCHMITT-TRIGGER
WAVESHAPES

output is in the high state (typically 3.3 V), the capacitor charges through R until
its voltage reaches V_{T+} (1.7 V), at which time the output switches low. Now the
capacitor discharges through R until its voltage reaches V_{T-} (0.9 V), when the
output once again switches to a high. The same type of analysis as that used in

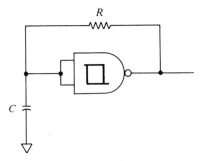

FIGURE 4.19   A SIMPLE OSCILLATOR USING A *NAND* SCHMITT TRIGGER

Section 2.12 applies here and is left to the reader. To summarize the performance, the time in the high state is $0.41RC$, while the time the output spends in
the low state is $0.64RC$. The frequency is then approximately $1/RC$. The timing
is not exact, nor is it predictable. This is due to the variation in the output high
voltage from one device to another. However, where a square-wave signal is
needed and accuracy is not essential, the circuit applies quite well.

4.3   FLIP FLOPS

The basic gates discussed above are examples of SSI (small-scale integration). SSI usually refers to ICs with an equivalent circuit complexity of 10 or
fewer gates. We now turn our attention to flip flops, an example of MSI
(medium-scale integration) which denotes ICs having an equivalent circuit complexity higher than 10 gates. The term LSI (large-scale integration) refers to ICs
having an equivalent circuit complexity higher than 100 gates.
      In simplest terms, a flip flop (FF) is a device whose output is controlled
through the input having the singular property that the output, once established
by the input, is maintained or remembered by the circuit even after the input is
removed.
      To understand the basic operation of flip flops, consider the interconnection
of *NAND* gates shown in Fig. 4.20. Let us assume both inputs high. If, for
example, Q is high, then both inputs to $N2$ are high and its output $\bar{Q}$ is low. Since
$\bar{Q}$ is connected as an input to $N1$, the low input causes Q to be maintained in the
high state. This is the basic latching or memory feature generic to all flip flops.
Had we originally assumed Q low, we would establish that the state of Q low and
$\bar{Q}$ high would be maintained. Thus one of the features of flip flops is that their
two outputs are logical complements of one another, therefore the usual labelling
by Q and $\bar{Q}$.

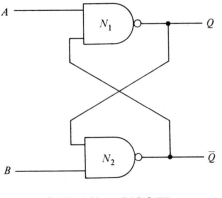

FIGURE 4.20   BASIC FF

From the previous discussion we see that with both inputs high, we can safely say that the outputs are maintained in their previous state. The condition of both inputs being high is then the unexcited state, that is, the inputs cause no change in the output. Consider now making the input to $N1$ low while maintaining B high. With a low input, the output of $N1$, Q, must be high irrespective of its previous state. Thus both inputs to $N2$ are now high, giving a low output at $\overline{Q}$. Thus the effect of making A low is to *set* Q to a high ($\overline{Q}$ assumes a low). Note that the circuit is symmetrical, that is, if we make B low while maintaining A high, $\overline{Q}$ is driven high and subsequently Q is driven low. The effect of making B low is to *reset* Q to a low. The inputs then accomplish the set and reset functions—A might be labelled as $\overline{S}$ and B as $\overline{R}$. The inversion is placed on both R and S since these functions are accomplished by applying a low to the respective inputs.

Should we make both inputs in Fig. 4.20 low at the same time, both outputs would be driven high. From a circuit standpoint, this is perfectly permissible —no damage to the circuit can result. However, from a logical standpoint, such a situation cannot be permitted since Q cannot be low or high at the same time as $\overline{Q}$. Therefore, no flip flop may be set and reset simultaneously.

Consider next the addition of two *NAND* gates to the input of the basic flip flop, as shown in Fig. 4.21. Consider the common input to N_3 and N_4, labelled clock, to be high. Applying a high to the set input (reset low) causes the output of N_3 to go low; this, in turn, causes the output of N_1 to go high. With the reset low, the output of N_4 is high; thus both inputs of N_2 are high and the output of N_2 must be low. The effect of applying a high to the set input is to cause a high at the Q output. Similarly, applying a high to the reset input while the set is low will eventually cause Q to be low. Thus this flip flop is an RS type while the clock input is high. As before, we do not allow the set and reset lines to be high together (the unexcited state for the inputs here is both low).

Suppose that the clock input is made low. The outputs of both N_3 and N_4 are forced high, irrespective of the R and S inputs, thus maintaining the previous

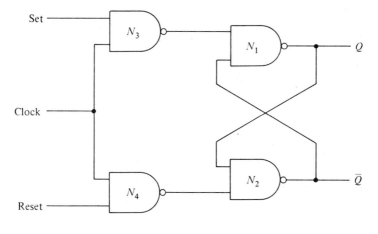

FIGURE 4.21   CLOCK RS FF

state of the flip flop. Thus the effect of the clock input is to determine when the R and S inputs can have an effect on the outputs. While the clock is low, the R and S inputs are disabled, that is, they cannot possibly cause a change in the output state. When the clock is made high, the R and S inputs are enabled, that is, they do have control of the output. This type of flip flop is termed a clocked RS FF.

The addition of the clock input allows synchronization. In a large digital system, many FFs and gates are all operating and delays in signals vary. With the clock we can enable specific FFs at specific times, thus forcing their outputs to change with a desired time relationship to one another.

Inputs which are enabled and disabled by the clock are called *synchronous*. Conversely, inputs which can affect the FF output without regard for the clock and therefore are not controlled by the clock are called *asynchronous*. If the two-input *NAND* gates N_1 and N_2 in Fig. 4.21 are replaced by three-input *NAND* gates as shown in Fig. 4.22, two additional inputs result. They are the preset (labelled Pr) and clear (labelled Cr) inputs. If maintained high, these inputs have no effect—operation of the FF is as discussed above. However, making Pr low forces the A output high (clock disabled, low). Thus Pr is an asynchronous input which, when activated, presents Q to be high. Similarly, making Cr low causes $\bar{Q}$ to go high or, conversely, Q to go low. So Cr is an asynchronous input which, when activated, clears Q to a low.

The logic symbols for the FFs discussed above are shown in Fig. 4.23. The basic FF of Fig. 4.20 is shown in Fig. 4.23(a); the FF of Fig. 4.21 is depicted in Fig. 4.23(b); the FF of Fig. 4.22 in Fig. 4.23(c). Note the small circle at the input—it is used to denote that the input is active low, that is, the input function is accomplished when the input is made low.

A data or D-type FF results when an inverter is added between the inputs of the clocked RS FF, as shown in Fig. 4.24. The clock terminal again acts to

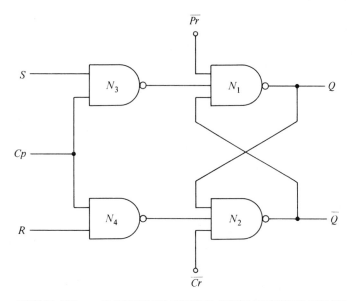

FIGURE 4.22   CLOCKED RS FF WITH ASYNCHRONOUS INPUTS

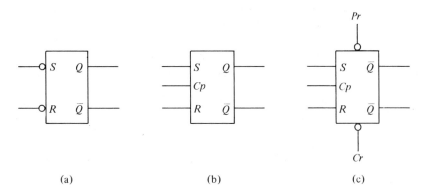

(a)   (b)   (c)

FIGURE 4.23   FF SYMBOLS: (a) BASIC FF; (b) CLOCKED RS FF; (c)
CLOCKED RS FF WITH ASYNCHRONOUS INPUTS

enable or disable the input. If the clock is high, and D is also high, the FF is
set—Q goes high. With the clock high and D low, the reset input is high and the
FF is reset. Thus, whatever the state at the D input, it is transferred to the Q
output while the clock is high. If the clock is low, the previous state is maintained
whether D is high or low. This type of FF, often called a latch, is very useful in
applications where a data line needs to be sampled occasionally and the sample
stored. The clock or latch enable, as it is often referred to, determines the instant
when the data line (D) is to be enabled. The output of the FF then assumes the

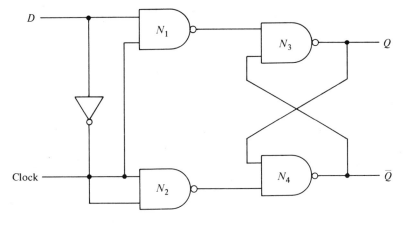

FIGURE 4.24   D-TYPE FLIP FLOP

same state as the data line until the enable is made low, at which time the previous data line state is stored in the FF and changes of state on the data line do not affect the Q output.

Yet another type of FF is shown in Fig. 4.25. This is the most versatile FF of all those discussed. Note that the *master-slave JK FF*, depicted in Fig. 4.25, actually consists of a cascade of two flip flops. The first, called the *master*, consists of gates N_1, N_2, N_3, and N_4; the second, called the *slave*, consists of gates N_5, N_6, N_7, and N_8. Each of the two FFs is a clocked RS-type. Note, however, that when the clock enables the master, it disables the slave due to the inverter in the slave clock line. Thus the input information from the J and K inputs is transferred into the master FF during the high of the clock with the slave inputs disabled. Once the clock goes low, the J and K inputs to the master are disabled and the R and S inputs of the slave are enabled, thus allowing the original input information of the J and K inputs to be now transferred to the outputs Q and $\overline{Q}$. The input gates, N_1 and N_2, have three inputs each, the third input being derived from the output of the slave. To see the effect of this feedback, consider both J and K high and the clock high. If Q is high, all three inputs to N_2 are high, and the master FF is reset (Q_M goes high). During this time the slave is disabled by the fact that its clock is low. When the clock goes low, the master is disabled and the slave R and S inputs are enabled. Since Q_M is connected as the reset input to the slave and it is high, the slave is reset forcing Q to go low. To summarize, with J and K high and Q starting out high, when the clock is brought high and then low, Q is forced low. From the circuit symmetry, it should be obvious that completely analogous operation would result had we assumed Q low initially ($\overline{Q}$ high)—that is, the set of the slave would be energized, thus causing the Q output to go high. Thus making J and K high together causes the FF output to *toggle* (change state) every time the clock returns low. If the clock input is a square-

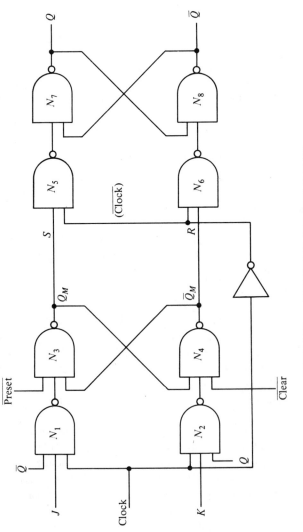

FIGURE 4.25 MASTER-SLAVE JK FLIP FLOP

wave with frequency f, the output will also be a square-wave, but with frequency $f/2$.

The master-slave FF can also be used as a conventional clocked RS FF, with the J input serving as the set and the K as the reset. Consider the outcome of making J high and K low. When the clock goes high, if Q is high, all three inputs to $N1$ are high, causing a low output which causes Q_M to go high. When the clock is made low, the slave is set by Q_M, resulting in Q high. Thus J high sets the FF. Note that had $\overline{Q}$ been in a low state originally, no states would have changed due to a high at J. But this is perfectly all right since Q would have been high already. From circuit symmetry, it should be apparent that analogous circuit action takes place when K is high with J low, resetting Q to low when the clock goes high and then low.

Note that asynchronous clear and preset inputs are also included in Fig. 4.25. A low input energizes these functions. They may not be brought low together and should be applied while the clock is low.

The FFs discussed so far are all level-triggered, that is, either a high level or a low level of the clock enables them. Another type of FF, called *edge-triggered*, is also available and extremely useful in certain applications. In the edge-triggered FF, the inputs are only enabled during the clock *transition* from high to low (negative-edge-triggered) or from low to high (positive-edge-triggered). Let us use the edge-triggered D FF shown in Fig. 4.26 as an example. This is one of two such FFs in a '74 IC. The FF is redrawn in Fig. 4.27 with the preset and clear inputs omitted for simplicity. Consider the D input high and the clock low. The outputs of N_2 and N_3 must be high; both inputs to N_4 are high causing its output to be low; this is an input to N_1, causing the output of N_1 to be high. Note that changing the D input while the clock is low does not affect the outputs since the N_2 and N_3 outputs are maintained high. When the clock is brought high, both inputs to N_2 are high, causing its output to go low. This forces Q high (the data is entered). No changes occur in N_3 and N_4. Now we can change the state of the D input, making it low. However, the low output of N_2 maintains a high at the output of N_3, so the Q output is unaffected. The D input only controls the Q output during the rising edge of the clock pulse.

If we start with Q low and the clock low, the states are the following: the outputs of N_2, N_3, and N_4 are high; the output of N_1 is low. When the clock goes high, all the inputs to N_3 are high, so its output goes low, causing $\overline{Q}$ to be high and, in turn, Q to be low. Once the output of N_3 has gone low, changing the D input has no effect since the output of N_4 is maintained in the high state by the output of N_3.

Note that the edge-triggered type clock input, when used in a JK FF, eliminates the need for a slave. The output feedback can be taken directly from the output of the master. The need for the master-slave arrangement was to prevent the race condition. This occurs if a level-triggered RS FF has feedback from the Q and $\overline{Q}$ output applied to the input. When the R and S inputs are made high, the

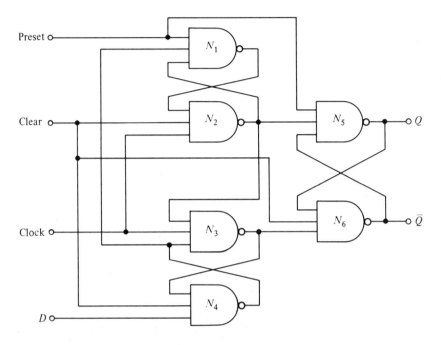

FIGURE 4.26   TYPE '74 EDGE-TRIGGERED D FLIP FLOP

FF would toggle but since the inputs would still be enabled by the clock, the FF would be free to toggle again and again. The output would oscillate. This was the purpose of the master-slave arrangement: to prevent the feedback from Q and $\overline{Q}$ from changing while the clock was high. If the FF is implemented in a similar manner to the D type discussed above, there is no longer the need for the slave. Edge-triggered JK FFs are also available, in both the positive-edge-triggered as well as the negative-edge-triggered versions.

The importance of the edge-triggered FF is that the inputs, be they JK or D, need only be steady during the clock transition. This is illustrated by comparing the waveshapes for a level-triggered master-slave FF (Fig. 4.28) to the waveshapes for an edge-triggered JK FF (Fig. 4.29). In Fig. 4.28, the FF is enabled when the clock is low. Thus after the second close pulse, since J is high, Q goes high. After the third clock pulse, Q does not change state since both J and K are low. When the clock goes low after the fourth clock pulse, since K is high, Q goes low. When the clock goes low after pulse 6, 7, and 8, since both J and K are high, Q toggles as shown. The output period during this time is twice that of the clock, so frequency division by 2 has been accomplished. However, note the following: should J go low even momentarily during the low clock time between pulses 6 and 7, the output Q would also go low for that period of time.

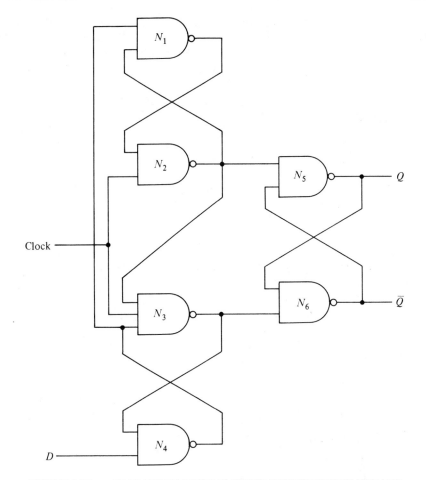

FIGURE 4.27   EDGE-TRIGGERED D FF (REDRAWN WITHOUT CLEAR AND
PRESET INPUTS)

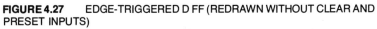
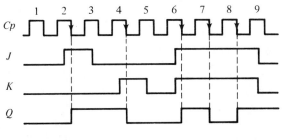

FIGURE 4.28   JK FLIP FLOP TIMING DIAGRAM

Consider now the timing diagram for a positive-edge-triggered JK FF shown in Fig. 4.29. Here the inputs are only enabled during the transition from low to high at the clock. Specifically, note that on the positive edge of pulse 7, with J and K both high, the output toggles. Even though J is shown as going low during pulse 7 and K is high with Q high, Q is not reset. Similarly, if K goes low momentarily with J high and Q low, Q is not set, as shown during pulse 8.

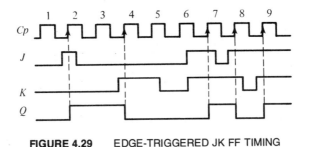

FIGURE 4.29   EDGE-TRIGGERED JK FF TIMING

The logic symbol for an edge-triggered JK FF is shown in Fig. 4.30. Note the little triangle at the clock input. This is used to specify an edge-triggered FF (the circle indicates that it is negative-edge-triggered).

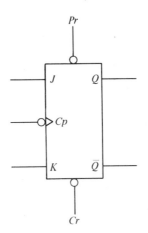

FIGURE 4.30   EDGE-TRIGGERED JK FLIP FLOP WITH PRESET AND CLEAR

In choosing between a master-slave and an edge-triggered JK FF, the criterion is not which is better in an absolute sense, but rather which one fits the application better. There are many applications in which the two are not inter-changeable, since the two do not function in exactly the same manner.

Once the appropriate FF is chosen for a specific application, it is essential to make certain that it is properly interfaced to the rest of the circuit. Typically, the fan-out from the FF outputs is 10 standard gate loads. The FF inputs present different loads depending on the particular circuit configuration used inside the FF. For example, for the '74 FF shown in Fig. 4.26, the D input presents one normalized load, that is, the loading on the driving gate connected to D is the same as if a single TTL input were connected; the loading on the preset and clock lines is equivalent to two TTL gates, while the clock is loaded by three TTL gates. If a logic diagram is available, the loading for a specific input of an FF can be determined by counting how many gates are connected to that particular input. Or using the specification of $I_{IL\ max}$ for that input, the TTL gate equivalent number is the $I_{IL\ max}$ divide by -1.6 mA, which is the maximum I_{IL} of a single TTL gate.

The notation used by manufacturers, although not completely standardized, is relatively easy to interpret. For example, in truth tables, either $\int$ or $\blacktriangle$ signifies that this is a positive-edge-triggered input ($\mathbf{\int}$ and $\blacktriangledown$ signify negative-edge-triggered). An X signifies that this input does not control the output—it is a "don't care" state and could be either low or high. For example, when a present input of a JK FF is made low, there may be X's in the same line under clock, J, and K, signifying that the Q output will be high irrespective of the clock, J, and K inputs.

4.4   COUNTERS

As we have already seen, when the J and K inputs are high and a square wave is applied to the clock input of a JK FF, the Q output toggles once for each clock cycle, thus providing frequency division by 2. This is the simplest counter—it counts to two: after one clock pulse its output is low, indicating a count of zero; after the second clock pulse its output is high, indicating a count of one.

This type of counting can be extended simply by cascading additional JK FFs. A four-bit binary counter is shown in Fig. 4.31. The scheme is extended to more than four bits simply by connecting the clock input of the next FF to the Q output of the previous one, with J and K permanently high. For the circuit shown, each FF toggles when its clock goes through a negative transition, from high to low. The timing diagram is shown in Fig. 4.32. The weighting of the binary outputs is as follows: Q_A—2^0, Q_B—2^1, Q_C—2^2, and Q_D—2^3. Thus, if we look at the outputs of the four FFs after the negative edge of the eleventh clock

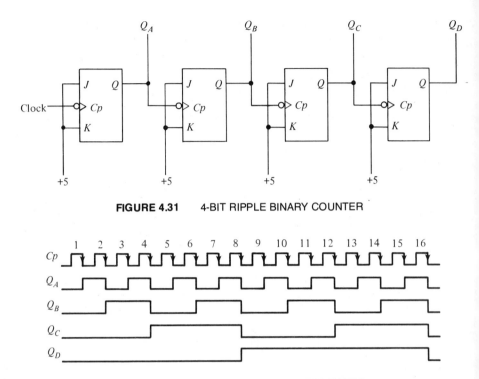

FIGURE 4.31   4-BIT RIPPLE BINARY COUNTER

FIGURE 4.32   4-BIT BINARY COUNTER TIMING

pulse, for example, we see that Q_A, Q_B, and Q_D are high with Q_C low. The corresponding decimal count is then $2^0 + 2^1 + 2^3 = 1 + 2 + 8 \doteq 11$. An example of a similar four-bit binary ripple counter is the '93, which, in addition, has reset inputs.

In many applications other counting schemes are used. An example of a BCD* (binary coded decimal) counter is shown in Fig. 4.33. The counting sequence here, although accomplished in a different manner than for the straight binary counter, is identical for the first nine clock pulses. When the tenth clock pulse goes through its negative transition, Q_D is reset and the counting sequence is repeated. The timing diagram for this counter is shown in Fig. 4.34. Note the following: if the connection between the Q_A output and the clock input of the second FF is broken, we have two separate counters: a divide-by-two counter formed by the input JK FF and a divide-by-five counter formed by applying the input to the clock of the second FF. To see the latter, consider the Q_A waveshape

---

*To illustrate the BCD representation of a number, consider the decimal number 73. In straight binary, this is 1001001. In BCD, it is 01010011, where the first group of four digits (7 in binary form) represents the 70 and the second group of four (3 in binary) represents the 3.

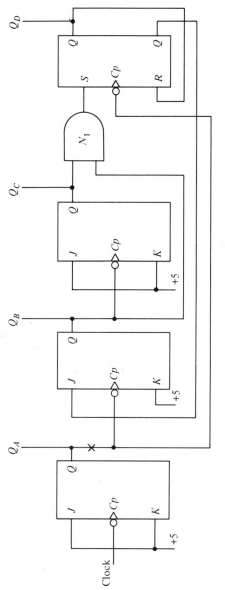

FIGURE 4.33 RIPPLE BCD DECADE COUNTER (7490)

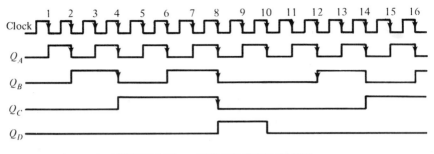

FIGURE 4.34   BCD COUNTER TIMING

in Fig. 4.34 to be the input. The three succeeding FFs count to five and then reset to zero. On the fifth negative transition of Q_A, the Q_D FF is reset. An example of such a counter which can be used to divide by two, five, or ten, depending on how it is connected, is the '90 counter. In addition to making the clock inputs of Q_A and Q_B available externally (all four FF outputs are also available externally), the '90 provides for asynchronous resetting of all four FFs to zero and for presetting to 9 (used in divide-by-10 applications) by presetting Q_A and Q_D.

Each BCD counter divides by 10. So if the Q_D output of one counter is connected to the Q_A clock input of another counter, the second counter is counting 10's, that is, when it reaches a count of 6, at least 60 clock pulses must have transpired. Succeeding decade counters connected in a similar manner will count 100's, 1000's, etc. Some obvious applications include frequency counters and DVMs.

The counters discussed above are termed *ripple* counters since the count, in a sense, ripples through the counter. Consider, for example, the sixth negative edge of the clock (refer to Fig. 4.34). As soon as this edge is applied, the gates inside the first FF switch, and shortly thereafter the Q_A output, toggle to a low. This negative transition is applied to the clock of the second FF, causing the gates of the second FF to switch state, toggling Q_B into a high state. Although the timing diagram shows the transition of Q_B to a high as coinciding with the transition to a low of the clock and Q_A, there is some delay between the input transition, the Q_A transition, and, eventually, the Q_B transition. The delay is only a few tens of nanoseconds, but in some applications it is intolerable. The *synchronous* counters avoid this delay by using a different counting scheme: each FF clock input is driven by the clock input simultaneously, causing the FF outputs to change simultaneously (synchronously). An example of a synchronous BCD and binary counters is shown in Figs. 4.35 and 4.36. The scheme illustrated makes use of multiple J and K input FFs. It functions as follows: all J inputs, for example, have to be high in order for the FF to be set (the different J inputs to the same FF are "*AND*ed" together as are the different K inputs of the same FF). The timing diagram of the synchronous counters is the same as that of the ripple

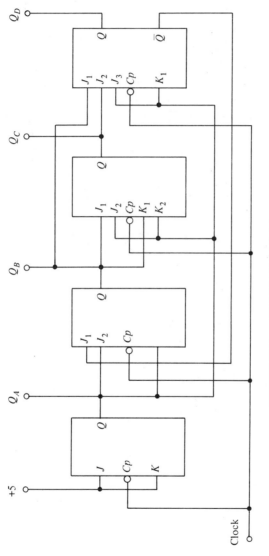

FIGURE 4.35 SYNCHRONOUS BCD COUNTER

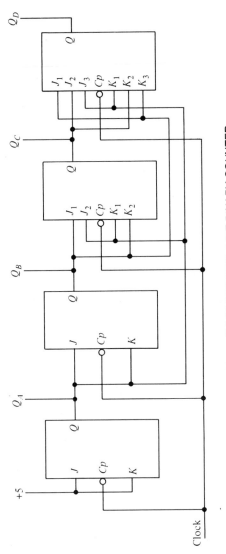

FIGURE 4.36 SYNCHRONOUS 4-BIT BINARY COUNTER

counter. The difference between the synchronous and ripple counters cannot be shown on that time scale; however, if the time scale at one of the negative input transitions is magnified, the difference becomes apparent. For example, consider the sixth negative clock transition as before. The waveshapes for the ripple and synchronous BCD counters (with an expanded time scale) are shown in Fig. 4.37. Note that in the ripple counter Q_A becomes low before Q_B goes high, while in the synchronous counter this occurs simultaneously.

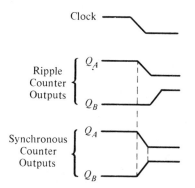

FIGURE 4.37   TIMING COMPARISON OF SYNCHRONOUS AND RIPPLE
COUNTERS

There is yet another minor difference between the two: in the ripple counter, there are counting spikes in the output waveshapes due to the asynchronous operation. These are not present in the synchronous counters.

As is the case with ripple counters, synchronous counters are available in different counting forms: decade (BCD), straight binary, etc. Another feature found on some counters is the *up-down* counting capability. Here two clock inputs are available. When the clock is applied to one, the counter simply counts up, as discussed above. However, when the clock is applied to the count-down input, through the use of proper gating (accomplished internally), the counter output count decreases by one for each clock pulse.

Another feature that is extremely useful in many applications is the capability of presetting the counter outputs to any value. Here, there are four data inputs available (one for each counter FF) with an additional load or enable input. When this load or preset enable input is energized, the data at the input is transferred to the counter outputs. This operation is asynchronous, that is, independent of the clock.

An example of a synchronous, presettable, up-down decade (BCD) counter on one chip is the '192, containing the equivalent of 55 TTL gates.

4.5   SHIFT REGISTERS

Quite similar in operation to counters, shift registers are used in many digital applications as digital delay elements and for data manipulation. An example of the latter is the conversion of serial data to parallel form, and vice versa.

An example of a serial shift register is shown in Fig. 4.38. Note that the first of the edge-triggered RS FFs is converted into a D FF by the inverter between the S and R inputs. Thus on the negative edge of the clock, the data input is entered into the first FF. On the next negative edge of the clock, this data is entered into the next FF and is available now at Q_B. At the same time (note that the FFs are clocked synchronously), the new data input is entered into the first FF. Thus on each succeeding negative edge of the clock pulse, data is shifted to the right, with new data accepted into the first FF and available at Q_A. This serial shift register thus provides the capability of converting serial data into parallel form. For example, assume that on four successive clock pulses, the data input corresponds to four bits of a serial digital word. (A digital word is comprised of a stated number of zeros and ones having significance when taken in a particular sequence—for example, 1001001 is a digital word of 7 bits corresponding to the decimal number 73.) When these four bits are shifted in over four clock pulses, the outputs Q_A through Q_D would contain the four-bit word, available in parallel (that is, all four bits simultaneously).

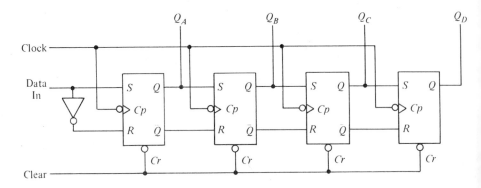

FIGURE 4.38   4-BIT SERIAL-IN PARALLEL-OUT SHIFT REGISTER

To see the operation of a serial-in/serial-out shift register (used for digital delay purposes), consider the same circuit shown in Fig. 4.38. The data is still applied serially to the data input. Only one output is taken, at Q_D. We can readily see that the data applied to the input is available at the Q_D output exactly four clock pulses later than it is applied. Thus, serial digital data (or signals) can be delayed by any integral multiple of the clock period simply by using a serial shift register having the appropriate number of bits. The scheme shown in Fig. 4.38 can be extended to any number of bits simply by cascading additional FFs. An example of an eight-bit serial shift register similar to the one in Fig. 4.38 is the 74164.

The reverse function is accomplished in a parallel shift register of the type shown in Fig. 4.39. Here, the input data may be entered in parallel or serial form, and shifted out serially. Consider first the shift/load input in a high state; the output of inverter I_2 is then low, thus forcing the outputs of NAND gates N_1 through N_8 high. This disables the parallel inputs—all preset and clear inputs are high (inactive). At the same time, with one input high, N_9 inverts the clock and applies it to the FFs which are clocked synchronously. Note that since the FFs are negative-edge-triggered and N_9 has inverted the clock, data transfer from the serial input will occur on the positive clock edge. With the shift/load input high (shift mode), the serial input is shifted and in this case delayed by four clock periods before it is available at the data out terminal. When the shift/load input is made low (load mode), the clock input is disabled (output of N_9 is high), but gates N_1 through N_8 are all enabled and functioning as inverters. If any of the parallel input lines is high, the output of the gate to which it is connected goes low and its FF is preset to a high. Conversely, if a parallel input is low, the output of the NAND gate to which it is connected is high, driving the clear input of its FF low, thus clearing the FF to low. In either case, the logic level at the parallel data input is transferred to the corresponding FF. Once the parallel data is loaded, the shift/load input is brought high, disabling the parallel load inputs and enabling the clock. After that, on each succeeding clock pulse the data in the FFs is shifted one position to the right and available at the data-out terminal. For example, assume that during the load mode, A, C, and D are high and B low. When the shift/load is brought high, the clock is enabled. After each successive clock pulse, the data-out terminal will be, in sequence, high, high low, and high. Thus a parallel word has been converted into serial form. This basic operation is available in a single TTL IC, the 74165, for example.

Other shift register configurations are available in single packages or can be implemented by a combination of shift registers. The basic operation is similar to that discussed in the two examples above.

4.6   MULTIPLEXERS

In many important digital applications, the need arises to multiplex or time-share a signal. An example is the output from clock and calculator chips. A number of typically seven-segment readouts need to be driven, each requiring seven inputs. If six readouts are used, this would require 42 lines just to drive the

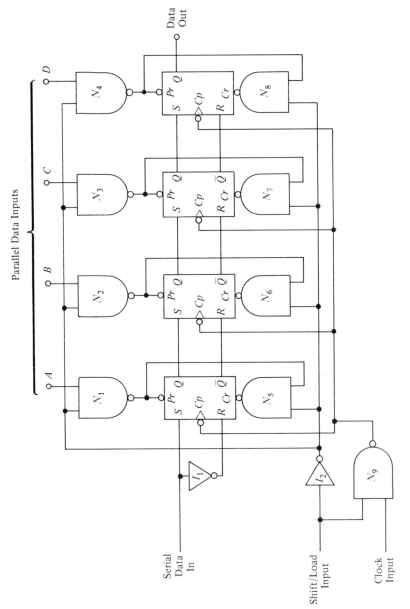

FIGURE 4.39 PARALLEL/SERIAL-IN SERIAL-OUT SHIFT REGISTER

readouts. If the segment outputs are multiplex, only seven lines are necessary for all the segments (which would be connected in parallel) together with six-digit enable lines, making a total of 13 lines as opposed to 42.

To see how a digital signal is multiplexed, consider the circuit in Fig. 4.40, which is half the dual four-line-to-one-line multiplexer in the 74153 TTL IC.

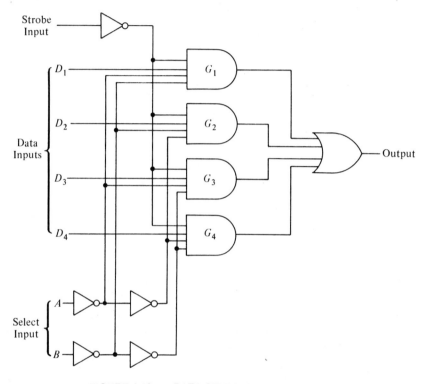

FIGURE 4.40   DATA SELECTOR/MULTIPLEXER

Let us first consider the operation of the *strobe* input. If the strobe input is high, one input of AND gates G_1 through G_4 is low, thus forcing all the AND gate outputs low and causing a low output. The strobe input in a high state disables all other inputs and gives a low output. If the strobe is low, the data inputs, together with the select inputs, determine the output. Note that with the strobe low, the output of G_1 is high only if D_1 is high at the same time that both A and B inputs are low. Thus when the select inputs are both low, the D_1 input is propogated to the output, since the outputs of G_2, G_3, and G_4 are all low. Thus the code of 00 at the A and B inputs selects the data at D_1 to be available at the output. In a similar manner, the select code at B and A of 01 selects the D_2 input; 10 selects the D_3 input; 11 selects the D_4 input.

The basic idea illustrated in Fig. 4.40 can be extended to handle any number of inputs. If three data select lines are used, eight data inputs can be multiplexed (the 74151, for example). Alternately, the output of one multiplexer can become the input to another. For example, if 4 four-line multiplexers are used as inputs to another four-line multiplexer, the proper drive to the data select lines will result in a single 16-line-to-1-line multiplexer. In such applications the strobe or enable input is used as an additional data select input by selecting a particular multiplexer.

To show the versatility of the multiplexer, we note that with the strobe enabled (low) and with the select lines fed from a two-bit binary counter, the multiplexer performs parallel to serial conversion (if the four-bit parallel word at inputs D_1 through D_2 is updated every four clock pulses).

4.7   DECODERS

In one sense, a decoder is just the opposite of a multiplexer (in fact the term *encoder* is sometimes used for a multiplexer). There are two basic types of decoders: code converters and demultiplexers. As an example of a code converter, consider the problem of converting an output of a BCD counter to drive a seven-segment LED (light emitting diode) readout. Since this presents as much of a problem in interfacing with the LED readout as in logic implementation, and also since such converters or decoders incorporate the driver circuit all on the same IC, we shall take up this problem in the section on interface ICs.

A four-line-to-sixteen-line decoder or demultiplexer is shown in Fig. 4.41. This circuit is available in a single IC package (74154) and clearly illustrates the circuit operation as well as complexity. If either or both enable inputs are made high, G goes low; since it is applied to all *NAND* gates N_1 through N_{16}, all 16 outputs are forced high. Thus both enable inputs must be low in order for the four-bit input to control the 16 output lines. It should be obvious that there are 16 combinations of the four input lines; these 16 combinations are decoded to provide a low at one of the 16 outputs. To see the operation, it is simplest to work backwards starting at the output. Consider that the output of N_{11} (labelled as 10) is low. If this is the case, all four of its input lines must be high. Therefore, A and C must be low with B and D high. An inspection of the inputs to the other 15 gates verifies that with this input condition, no other gates have a low output. If we consider the four-bit input to be a binary word corresponding to a decimal number with D the most significant bit, the input of 1010 is $8 + 2 = 10$. Thus the output of N_{11} is labelled 10. Each of the other output numbers corresponds to the binary input code required to give a low at that output.

An example might illustrate one application for multiplexers and decoders. Let us assume that a 16-bit binary word corresponding to some intelligent information (for example, an output of a computer) needs to be transmitted over some

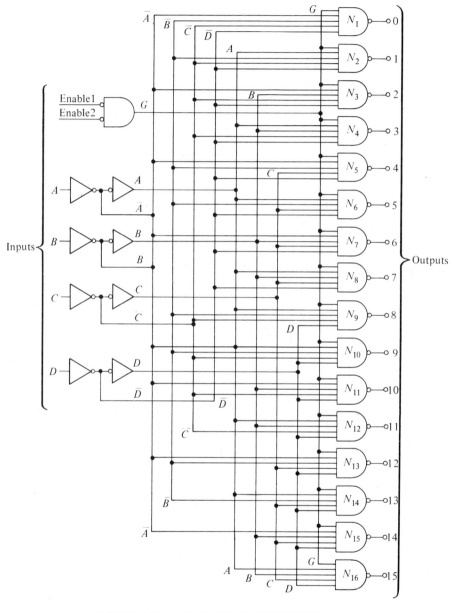

FIGURE 4.41   4-LINE-TO-16-LINE DECODER

distance and used for some remote control function. If this word changes periodi-
cally (at predictable intervals), we normally would have to connect 16 wires for
the 16 bits, plus a ground return wire, or a total of 17 wires. However, if we take

the 16 parallel bits and serialize them, by using either a parallel-in/serial-out shift register or a 16-line-to-1-line multiplexer, we have reduced the 16 lines to one. Now the number of lines necessary is one for the data line, one for the synchronizing pulse or clock, and one for the ground return—a total of three lines. At the other end of the lines, we need to reconstruct the serial stream into parallel form. Again, two methods are available: either by clocking serial data into a serial-in/parallel-out 16-bit shift register, or by feeding the clock into a four-bit binary counter whose outputs are applied to the inputs of a 4-line-to-16-line decoder. The 16 parallel bits are reconstructed by *AND*ing the serial data line with the inverted 4-lines-to-16-line decoder outputs. (Note that one of the three lines can be eliminated by inserting a synch pulse between adjacent words; however, in that scheme the second decoding method outlined above cannot be used.)

REVIEW QUESTIONS

4.1 In comparison with regular TTL, list the characteristics of the low-power TTL. Repeat for the high-speed TTL, Schottky TTL and low-power Schottky TTL.

4.2 What is the typical input current for a 7400 gate when the input is low (below 0.4 V)? Repeat for 74L00, 74H00, 74S00 and 74LS00.

4.3 What is the maximum input current for a 7400 gate when the input is high (above 2 V)? Repeat for 74L00, 74H00, 74S00 and 74LS00.

4.4 What are the maximum low and minimum high voltages for the different TTL families?

4.5 What is the unallowed range of input voltages for the standard TTL family? If the input is not brought rapidly through this unallowed range of voltages, what are the consequences as observed at the output?

4.6 What is the significance of the number listed as the fan-out for a given logic family?

4.7 List examples when different TTL families might be mixed (i.e. when more than one TTL family might be used in the same design), and give reasons why. (Assume that the standard TTL is always lower in cost than any other family type.)

4.8 What is the fan-out for a low-power Schottky driving low-power TTL gates? Why would such a configuration come about?

4.9 What is the fan-out of a TTL driving a low-power TTL, and why would such a configuration be used?

4.10 What is the fan-out for a high-speed TTL driving a standard TTL, and why would such a configuration result?

4.11   What is the fanout for a high-speed TTL driving a standard TTL, and why would such a configuration result?

4.12   Obtain a circuit diagram for a two-input TTL *NOR* gate and make a determination of the maximum low-level as well as the minimum high-level input voltage. (Follow the treatment for the *NAND* gate in the text.)

4.13   If an inverter is implemented using a *NAND* gate as shown in Fig. 4.14, what is the loading on the output of the gate that provides the signal A for the circuits in (a) and (b)?

4.14   The method shown in Fig. 4.14 (b) is preferable to that in (a). Why?

4.15   Replace the two *NAND* gates in Fig. 4.20 by two 2-input *NOR* gates. What kind of FF results?

4.16   Why are clocked FFs more widely used than unclocked ones?

4.17   What is the difference between synchronous and asynchronous inputs to an FF?

4.18   Some clocked FFs are edge-triggered and some are level-triggered. What is the difference between them?

4.19   What is an RS FF?

4.20   What is a D FF?

4.21   What is a JK FF?

4.22   In what way can a JK FF be used to perform the same function as an RS FF?

4.23   How can a toggle FF be implemented using a JK FF?

4.24   Explain the idea and need behind the master-slave JK FF.

4.25   What is the loading factor (number of TTL loads) on the clock input for the master-slave JK FF as shown in Fig. 4.25?

4.26   Specify the loading factor for each of the inputs to the edge-triggered D FF shown in Fig. 4.27.

4.27   Two types of counters are available: ripple and synchronous. What are the differences between these two types?

4.28   If a 1 MHz square wave is fed to the clock input of the synchronous 4-bit ripple counter shown in Fig. 4.36, what would be the frequencies available at the four outputs (Q_A, Q_B, Q_C, and Q_D)? Would any of the outputs have a 0.5 duty cycle (equal high and low times)?

4.29   Using the 7490 decade counter, implement a circuit to provide a 1 kHz square wave with 0.5 duty cycle from an input of 1 MHz.

4.30   Implement a circuit to provide a delay equal to four periods of the system clock. (*Hint*: Use the shift register shown in Fig. 4.40.)

4.31   Repeat Question 4.30 using the shift register shown in Fig. 4.39.

4.32   Using decade counters and the multiplexer shown in Fig. 4.41, implement a system that has a 10 MHz input and provides a programmed output, controlled by a 2-bit code as follows:

A	B	OUTPUT FREQUENCY
0	0	10 MHz
0	1	1 MHz
1	0	100 kHz
1	1	10 kHz

4.33   Implement a system to multiplex four digital input signals (available simultaneously on four parallel lines) to a single signal to be transmitted over a pair of wires.

4.34   Implement a receiver for Question 4.33 that will separate the four signals and provide four separate outputs from a single input.

5

CMOS

CMOS (complementary metal-oxides-semiconductor transistor) digital ICs have undergone such profound changes since their introduction that they now represent a viable logic family competitive with TTL. The basic families are the 4000 line produced by most IC manufacturers and the newer 54C/74C line which is a pin-for-pin direct replacement for the TTL line. Because it has both an ever wider variety of logic building blocks available and the obvious advantage of having lower power requirements than TTL, the CMOS logic line is assuming a major role in all low-power, low-speed applications.

It is essential that the IC user become familiar with the specific characteristics—the strong points as well as the limitations—of CMOS ICs in order to make the right choice of logic family for a particular application. It is only by thoroughly comparing logic families and evaluating them against the system requirements that one can make the right choice.

5.1   BASIC OPERATION AND CHARACTERISTICS

Perhaps the most unique characteristic of CMOS is that only two devices are necessary to implement all logic functions; no resistors are required. The two devices are N- and P-channel enhancement MOS transistors, thus the name complementary since the N- and P-channel devices are complementary as well as connected in a complementary configuration. The construction of the P-channel MOS transistor (actually a field-effect transistor) is shown in Fig. 5.1. As the cross-sectional diagram shows, the drain and source terminals are formed by

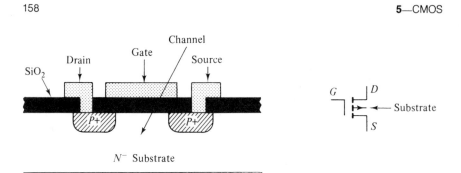

FIGURE 5.1   *P* -CHANNEL ENHANCEMENT-MODE MOSFET STRUCTURE
AND SYMBOL

heavily doped *P*-type wells in the lightly doped *N*-type substrate. The gate
terminal exercises control over conduction between the drain and source: if the
gate is made sufficiently negative, a positive charge is induced into the channel
(*N*-type region between drain and source) forming a low impedance path between
the drain and source. Under these conditions, the *P*-type MOSFET is on and has
a very low resistance as well as voltage drop from drain to source. The oxide
(SiO$_2$) layer, typically 100 Å in thickness, acts as a dielectric, that is, it forms a
capacitor between the metal gate and the channel. This action attracts additional
holes into the channel causing the channel to be *P*-type. If, on the other hand, the
gate has zero bias, no *P*-type channel is induced and an extremely high resistance
exists between the drain and source—the MOSFET is off.

The cross-sectional structure of an *N*-type MOSFET is shown in Fig. 5.2.
The drain and source are connected to highly doped *N*-type wells in a lightly
doped *P*-type substrate. This structure is incorporated on the same substrate
(lightly doped *N*-type) as the *P*-channel MOSFET. The metal gate is again

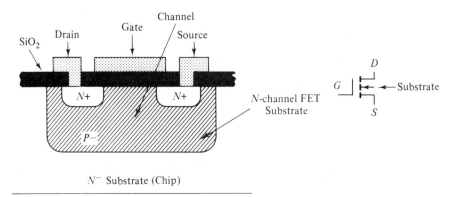

FIGURE 5.2   *N* -CHANNEL ENHANCEMENT-MODE MOSFET STRUCTURE
AND SYMBOL

isolated by the oxide layer. When a positive bias is applied to the gate, a negative charge is induced into the channel, providing for a low resistance path between the drain and source—the MOSFET is on. However, if zero bias is applied to the gate, no N-type channel is induced, and a very high resistance between the drain and source is realized—the MOSFET is off.

The minimum gate bias voltage (positive with respect to the source for an N-channel, negative for a P-channel) causes the MOSFET to pass a specified drain current, called the *threshold* voltage, abbreviated V_T. Typical transfer characteristics for N- and P-channel MOSFETs are shown in Fig. 5.3. Note that once the gate bias exceeds the threshold voltage, the drain current is proportional to the square of the gate voltage.

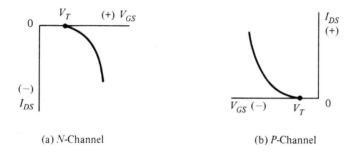

(a) N-Channel   (b) P-Channel

FIGURE 5.3   MOSFET TRANSFER CHARACTERISTICS

Consider the inverter formed by interconnecting an N- and a P-channel MOSFET, as shown in Fig. 5.4. If V_{SS} is connected to ground and $V_{DD} = V_{cc}$, a positive supply voltage, we have the following operation: with V_i at or near

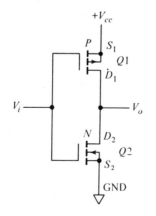

FIGURE 5.4   CMOS INVERTER

ground, the gate of $Q1$ is negative with respect to the source, and assuming V_{cc} greater than V_T, $Q1$ is on. At the same time, the gate-source voltage of $Q2$ is below V_T, thus causing $Q2$ to be off. With the high resistance of $Q2$, the P-channel MOSFET saturates $(V_{DS} \sim 0)$, and V_o is approximately equal to V_{cc}. Thus a low input causes a high output. Now suppose that the input is high, near V_{cc}: the gate of $Q2$ is now positive with respect to its source, and again assuming the input to be in excess of V_T, $Q2$ is on. At the same time, the gate-source voltage of $Q1$ is almost zero, thus causing $Q1$ to be off. With the high resistance of $Q1$ in series with it, $Q2$ saturates, and the output is essentially zero volts. Thus a high input causes a low output—the inverter function has been established.

CMOS devices typically have threshold voltages of 2 V. The resulting transfer characteristics for three different supply voltages are shown in Fig. 5.5.

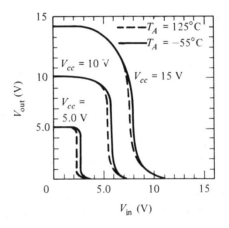

FIGURE 5.5   TYPICAL INVERTER TRANSFER CHARACTERISTICS

When the input is between 0 and approximately 2 V, $Q1$ is on and $Q2$ is off. Also, with the input between V_{cc} and $V_{cc} - 2$ V, $Q2$ is on with $Q1$ off. However, with an input between 2 V and $V_{cc} - 2$ V, both $Q1$ and $Q2$ are on. For digital applications, it is essential that the low-to-high or high-to-low input transition occurs as quickly as possible to minimize the time during which both the N- and P-channel devices are in conduction, thus minimizing power dissipation.

Note that because applying either a high or a low gives us one device on while the other is off, the net current drain from the power supply under static operation is almost insignificant. It is only during switching that significant amounts of current are drawn from the supply. With V_{cc} greater than $2V_T$, the supply current waveshape due to overlap in conduction is shown in Fig. 5.6. The resulting supply power drain is obviously directly proportional to the slope of the input waveshape, the power supply voltage, and the frequency.

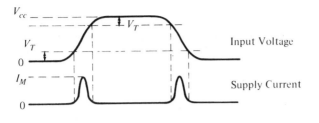

FIGURE 5.6   INVERTER SUPPLY CURRENT WAVEFORM

   The other part of the unloaded power dissipation of a CMOS inverter (or any other gate, for that matter) is the power required to charge the effective capacitance at the input. Note that the gate structure forms a capacitor, and in order to charge and discharge this capacitance, current must be supplied. The net effect of the gate capacitance as well as the current drawn due to the overlap in conduction can be modelled by a net effective capacitance C_{PD}. In one cycle of the input, this capacitance must be charged and discharged: the energy stored in the capacitor charged to V_{cc} is $\frac{1}{2}C_{PD}V_{cc}^2$. Thus, since the capacitor is both discharged and charged for each cycle, the net energy supplied (per cycle) must be twice the amount stored, or $C_{PD}V_{cc}^2$. The power is ascertained by determining the energy per unit time or by multiplying the energy per cycle by the time of one cycle: $P_D = C_{PD}V_{cc}^2 f$ (where f is the frequency of the input).
   To determine the total dissipation per inverter, we need only include the effect of driving the load capacitance C_L. Thus the total power dissipation is $(C_{PD} + C_L)V_{cc}^2 f$. The effective power dissipation capacitance C_{PD} is listed by the manufacturer, and depends on the particular device. The load capacitance may be determined by using the value of input capacitance (also listed by the manufacturer) and multiplying it by the number of such gates connected to the output. For example, the manufacturer lists $C_{PD} = 12$ pF and $C_{IN} = 6$ pF for each inverter in the 74C04 hex inverter. Thus we can estimate the power dissipation per inverter, if loaded by two other inverters operating from a 5 V supply with an input frequency of 10 kHz, to be approximately 6 μW. If the input frequency is 1 MHz, the power dissipation is 0.6 mW. If the power supply is raised to 15 V, the power dissipation is 54 μW at 10 kHz and 5.4 mW at 1 MHz.
   From the above discussion we can conclude that to minimize power dissipation, a low power-supply voltage should be used. However, the choice of power-supply voltage is not quite that simple. If we consider the switching characteristics of CMOS, we note that the propagation delay (time to cause the output of the gate to switch from a high to a low or vice versa) is a function of the supply voltage. That is, the charging of the effective internal capacitance is accomplished by passing a constant current through the conducting MOSFET —this current is a function of the geometry of the MOSFET and the supply voltage: the higher supply voltage, the higher the charging current and the lower

the propagation delay. A typical set of characteristics is shown in Fig. 5.7. Note that the decrease in propagation delay is proportional to the increase in the supply voltage. Also, for a given supply voltage, the larger the load capacitance (the sum of the input capacitances of gates connected to the output), the longer the propagation delay.

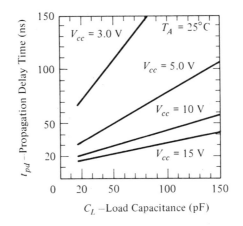

FIGURE 5.7   TYPICAL CMOS PROPAGATION DELAY CHARACTERISTICS

The choice of power supply voltage for a particular application is dictated by the compromise between propagation delay (which determines the maximum signal frequency) and power dissipation. It is not possible to optimize both—that is, to minimize both the power dissipation and the propagation delay. A good approach is to make the propagation delay approximately one-tenth (or smaller) of the shortest signal period (highest signal frequency) and determine the power supply voltage accordingly, accepting the resulting power dissipation.

The structure of the CMOS inverter is shown in Fig. 5.8; this corresponds to the schematic of Fig. 5.4. Although not indicated in the figure, the chip substrate, which is the channel of the P MOSFET, is connected to its drain which is at $+V_{cc}$. Similarly, the $P-$ substrate, which is the channel of the N MOSFET, is connected to its source, which is at ground. The resulting structure has a parasitic SCR (silicon-controlled rectifier) between the supply and ground. To see this, consider the two-transistor representation of the SCR as shown in Fig. 5.9. The NPN transistor is formed by the source of $Q2$, the $P-$ substrate, and the $N-$ substrate; the PNP transistor is formed by the source of $Q1$, the $N-$ substrate, and the $P-$ substrate. The relatively low bulk resistances of the two substrates, R_N and R_P, are in parallel with the base-emitter junctions. This has the effect of degrading the transistor β, which in turn degrades the parasitic SCR—in other words, we have a low-quality SCR. This means that the gate current required to

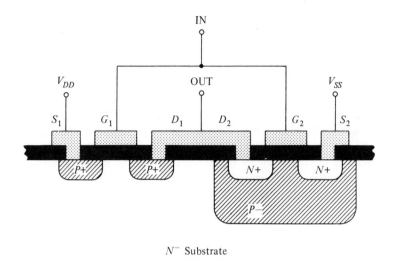

FIGURE 5.8   CMOS INVERTER STRUCTURE

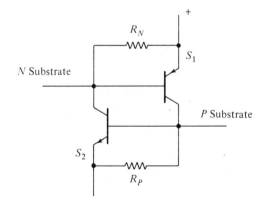

FIGURE 5.9   EQUIVALENT PARASITIC SCR FORMED IN CMOS INVERTER

turn on the SCR is quite large. Under normal operation of the inverter, such a
large current is highly unlikely, so the parasitic SCR never turns on. However, it
is not impossible for the SCR to be activated. Should the SCR fire, it places an
extremely low effective resistance across the power supply and draws significant
current. Under these conditions, the output is no longer controlled by the input
(the gate is malfunctioning); the resulting power dissipation may permanently
damage the IC. Assuming that the gate has not been damaged, the only way to
turn off the parasitic SCR is to disconnect the power supply. In order to prevent

this undesired circuit action, it is essential that the power supply voltage be on prior to any input signal. This is especially important to remember in bench testing where the input signal may be applied from a pulse generator: the power supply must be turned on before the pulse generator is connected.

The fan-out of CMOS driving CMOS is essentially infinite under static conditions since the gate draws negligible current. It is only during switching that the output of a gate has to supply (or sink) current to charge the effective load capacitance. However, the input capacitance of CMOS gates is not standardized. It depends on the physical dimension of the gate as well as the circuit configuration (type of gate). The fan-out then is limited by the signal frequency (the larger the load capacitance, the longer the propagation delay) and the maximum power dissipation of the particular device and/or IC package.

5.2   SPECIAL CONSIDERATIONS

All CMOS devices incorporate some sort of protection. This is essential since the thickness of the oxide layer is so small that even static discharge could break down the oxide and destroy the device. A typical protection scheme is shown in Fig. 5.10. Diode D_3 is formed by the two substrates and only becomes activated should the power supply be reversed. Input protection is provided by D_1 and D_2: should the input become more positive than V_{cc}, D_1 conducts and clamps the input at $V_{cc} + 0.6$. The gates are further protected by the series resistance R (typically 500 Ω), which limits any input current. Should the input become negative, D_2 conducts and clamps the gates to -0.6 V. Alternately, D_2

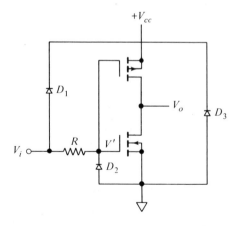

FIGURE 5.10   CMOS INPUT PROTECTION

can be connected between the input and ground, with basically the same operation.

Although all gates have this internal protection, it is still a good idea to exercise caution in order to prevent static discharge through the gates of the devices. A few simple rules should be observed:

1. Always store CMOS devices in a low resistance manner.
2. Never insert or remove a CMOS device while power is on.
3. Always reference unused inputs to either $+V_{cc}$ or ground (depending on the desired logic).

The first of these rules means that the leads of any CMOS devices should be electrically interconnected with minimum resistance. Typically, this is accomplished by inserting the IC pins into a special conductive foam or styrofoam covered by aluminum foil (the manufacturers usually supply the devices in an anti-static plastic tube or conductive foam). It is a good idea to use this method whenever the IC is not in the circuit. When transporting ICs across a room, do not carry them by hand or in a plastic container, especially if the room is carpeted (carpeting tends to build up tremendous static). The ICs should be either in the conductive foam or in aluminum foil or should be carried in a tin tray with the leads in contact with the tin. Similarly, do not store unprotected devices in plastic drawers—either line the plastic drawers with aluminum foil or have the leads inserted into the conductive foam.

While soldering the leads of any CMOS device, or while a CMOS device is in the circuit and other components are being soldered, make certain that the tip of the soldering iron (30 W max.) is properly grounded.

While the above precautions sometimes may be disregarded with impunity, nevertheless it is worthwhile to observe these few simple rules to insure the integrity of CMOS/ICs.

5.3   CMOS GATES

The CMOS family of ICs is complete to the extent that inverters and *NAND* and *NOR* gates as well as flip flops, counters, etc., are available. In fact, with the introduction of the 54/74C line, most of the functions available in TTL are duplicated in CMOS.

Consider the circuit diagram for the two-input *NAND* gate, shown in Fig. 5.11. Note that the basic complementary arrangement of the inverter is maintained. A high output, essentially V_{cc}, results if either input or both inputs are low, since $Q1$ or $Q2$ or both are on, with both $Q3$ and $Q4$ off. If both inputs are

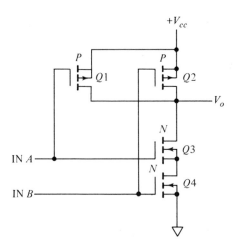

FIGURE 5.11   CMOS TWO-INPUT *NAND* GATE

high, $Q1$ and $Q2$ are both off, while $Q3$ and $Q4$ are both on, causing a low output (essentially ground). The function implemented is a positive logic *NAND*. This scheme obviously can be extended to any number of inputs simply by adding one N- and one P-channel MOSFET for each additional input—the P-channel in parallel with $Q1$ and $Q2$ and the N-channel in series with $Q3$ and $Q4$. Thus a *NAND* gate with any number of inputs can be constructed.

A two-input positive logic *NOR* gate is shown in Fig. 5.12. Here the P-channel MOSFETs are connected in series, while the N-channel devices are

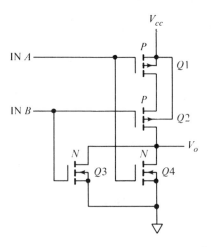

FIGURE 5.12   CMOS TWO-INPUT *NOR* GATE

connected in parallel. If either or both inputs are high, $Q3$ or $Q4$ or both conduct, and the output is low. If both inputs are low, $Q3$ and $Q4$ are both off, while $Q1$ and $Q2$ are both on, thus giving a high output. This is then the *NOR* operation. Again the scheme can be extended to provide a *NOR* gate with any number of desired inputs simply by adding a complementary pair of MOSFETs for each additional input: the additional *P*-channel MOSFETs would be placed in series with $Q1$ and $Q2$, while the *N*-channel devices would go in parallel with $Q3$ and $Q4$.

Although no protection circuitry is shown in Figs. 5.11 and 5.12, all CMOS gates have protective diodes and the series resistor incorporated on the chip.

5.4   TRANSMISSION GATE

The transmission gate has no counterpart in bipolar (TTL) circuitry. Besides allowing for different implementation of flip flops and other digital functions, it is very useful in analog applications.

The basic transmission gate is shown in Fig. 5.13. It consists of an inverter formed by $Q1$ and $Q2$, *P*- and *N*-channel MOSFET as before. In addition, two complementary MOSFETs are included with uncommitted source and drain terminals. Note that the control input signal is applied directly to the gate of $Q3$, while the gate of $Q4$ is driven by the control input after it has been inverted. Thus $Q3$ and $Q4$ are on together while the control input is high, and are off together while the control input is low. Consider the effect of this action on the input connected to $Q3$ and $Q4$: if the control input is high, $Q3$ and $Q4$ are on (the

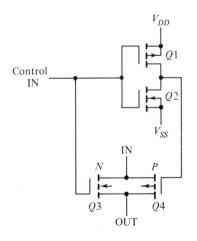

FIGURE 5.13   CMOS TRANSMISSION GATE

parallel connection reduces the *on*-resistance), and the input is transmitted to the output. If, on the other hand, the control input is low, both $Q3$ and $Q4$ are off (in a high resistance state), and the input is effectively isolated from the output. Thus a high on the control line transmits the input to the output, while a low isolates the output from the input.

For digital applications, $V_{DD} = V_{cc}$ (3 to 15 V), and V_{SS} is connected to ground. By applying a control signal from another CMOS gate connected to the same supply, another digital signal can be switched on and off through $Q3$ and $Q4$. Since these MOSFETs are symmetrical, the signal input and output are interchangeable, that is, the transmission gate passes signals in either direction. We shall examine a typical digital application in the next section.

Another important application for the transmission gate is the digitally controlled analog switch: a switch which can pass analog signals. For these applications it is essential that the digital control signal, the power supply voltages, and the range of analog signals be properly coordinated. By using the same supply for the driving CMOS gate as for the transmission gate, the proper relationship between the control input and the power supply is automatically established. For positive turn-on and turn-off of the transmission gate, the control input in the high state should be equal to V_{DD}, and in the low state, equal to V_{SS}. This insures the proper operation of the $Q1$-$Q2$ inverter. For proper operation of the switch ($Q3$ and $Q4$), the maximum analog voltages at the input or output must be more positive than V , but less than V_{DD}. For example, if the analog signal is between 0 and +5 V, the transmission gate can be operated from any supply voltage (V_{DD}) between +5 and +15 V with V_{SS} at ground. If, on the other hand, the analog input is between ±7.5 V, V_{DD} should be +7.5 V with V_{SS} equal to −7.5 V.

The CD4016 chip contains four such transmission gates or analog switches, and provides typically 200 Ω *on*-resistance and leakage of 100 pA in the *off* state. For applications requiring lower *on*-resistance and one more constant over the analog input range, the CD4066, which is a pin-for-pin replacement for the CD4016, should be used. In either case, using the CD4016 or the CD4066, the *on*-resistance can be lowered simply by parallelling two or more transmission gates. This is accomplished simply by connecting all the control inputs, all the analog inputs, and all the analog outputs together. (Remember that this will increase the capacitive loading on the digital drive to the control input, as well as increase the *off*-state leakage.)

5.5   FLIP FLOPS

Basically the same types of flip flops are available in CMOS as in TTL. Logical operation is quite similar, and the same precautions apply—most importantly, the clock transition for all edge-triggered FFs should be fast to prevent multiple triggering. However, the internal implementation of CMOS FFs is accomplished in a somewhat different fashion because of the transmission gate used.

The logic diagram for a type D FF (74C74) is shown in Fig. 5.14. For a comparison with the TTL version, see Fig. 4.28. The logic block used for the transmission gates has the following convention: the gates labelled T are on when the clock is high and off when the clock is low. Those labelled $\overline{T}$ are on when the clock is low and off when the clock is high. This is accomplished by using the clock as the control input for the T gates, while using the inverted clock as the control input for the $\overline{T}$ gates. To see the basic operation, consider the asynchronous inputs, clear and preset, to be inactive, that is, both high.

There are two flip flops: the master formed by N_1 and N_2, and the slave formed by N_3 and N_4. If the clock is low, the following is true: transmission gates T_1, T_4, and T_5 are on, with T_2, T_3, and T_6 off. The Q and $\overline{Q}$ outputs are taken from the slave FF through T_4 and T_5. The data input D is entered into the master through T_1, so that the output of N_1 is D. When the clock goes high, T_1 turns off, causing the data input to be now disabled and no longer in control of the FF. Thus only the information on the data line present during the positive clock edge is entered. T_3 and T_6 are now on, applying Q to I_4 and $\overline{Q}$ to I_5 (note that T_4 and T_5 are off). Thus $Q = D$ and $\overline{Q} = \overline{D}$. At the same time, the data is transferred from the master into the slave through T_3. The complete cycle is as follows: data is entered into the master and transferred to the slave as well as the output on the positive edge of the clock; data is disabled at any other time (since no path exists from D to either Q or $\overline{Q}$).

The asynchronous clear and preset inputs are tied together, and each has only one line brought out externally (the diagram shows two inputs for each only for simplicity of drawing). To understand the asynchronous operation, first consider the clock low. If preset is made active (low), the output of N_4 is made high, and with T_4 on, the output of I_4 is forced low. Also, with clear high, both inputs to N_3 are high, so its output is driven low, and with T_5 on, the Q output goes high. Had we started with the clock in the high state, the preset function would act on the master FF, forcing the output of N_1 high, which with T_3 on would cause $\overline{Q}$ to go low. At the same time, since both inputs to N_2 are high, its output is low, and with T_6 on, Q is driven high. Note that whatever the state of the clock, low or high (but not during the low-to-high transition), when the preset input is brought low, the Q output becomes high.

Since the two FFs are symmetrical and the clear and preset inputs are applied symmetrically, the action of clear is the same for the $\overline{Q}$ output as that of preset for the Q output. Therefore, making clear low causes Q low and $\overline{Q}$ high, irrespective of the clock (as long as it is not during its low-to-high transition when the D input is active).

When compared to the operation of the TTL 7474, we see that the internal CMOS implementation is different, while the terminal characteristics are identical. Thus in using CMOS flip flops, counters, decoders, etc., the same rules

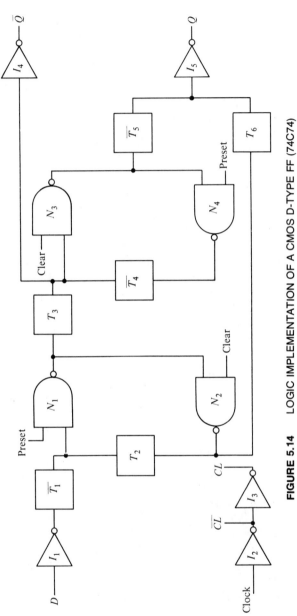

FIGURE 5.14 LOGIC IMPLEMENTATION OF A CMOS D-TYPE FF (74C74)

apply as for using TTL, even though the internal operation is accomplished in a slightly different manner due to the use of transmission gates.

5.6   CMOS IN LINEAR APPLICATIONS

Although CMOS gates are primarily used in digital applications, they can be used to advantage in some linear applications. For example, examination of the voltage transfer characteristics of Fig. 5.5 shows that if the input is biased at near $\frac{1}{2}V_{cc}$ and a small ac signal applied, a large output ac signal results (the slope of the transfer curve near the bias point is the voltage gain). This property is used to advantage in the small-signal ac amplifier of Fig. 5.15. Here, three inverters are cascaded to give a large open-loop amplification (any odd number of inverters can be used). The configuration has self-bias: the connection of R_2 (typically 1 MΩ) biases the input at approximately $\frac{1}{2}V_{cc}$. The gain is set by the ratio of resistors, much like OP AMPs. The lower 3 dB frequency is set by the input time constant, R_1C, while the upper 3 dB frequency is a function of the inverters (typically in excess of 1 MHz).

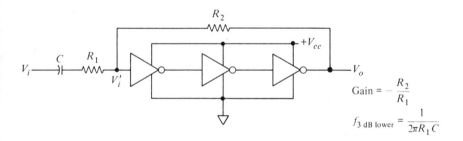

$$\text{Gain} = -\frac{R_2}{R_1}$$

$$f_{3\text{ dB lower}} = \frac{1}{2\pi R_1 C}$$

FIGURE 5.15   74C04 CMOS INVERTER AS A LINEAR AC AMPLIFIER

Another feature of CMOS circuits is that they operate with the output anywhere between the supplies; i.e. depending on the input, the output can be driven to within a few millivolts of either ground or V_{cc}. In OP AMP applications with a low power-supply voltage, CMOS circuits extend the range of output voltages when used as a post amplifier. A typical application is shown in Fig. 5.16. With the post amplification provided by the *NOR* gates (actually used as inverters), the output voltage range is between zero and +5 V while the same OP AMP without post amplification allows for an output between zero and 3.6 V (the output saturation voltage with $V_{cc}=5$ V).

In all linear applications, the key to successful use of CMOS gates is to make

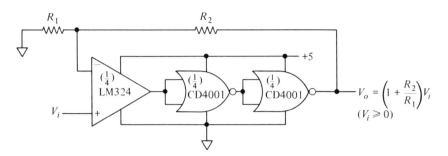

FIGURE 5.16   CMOS OUTPUT BUFFER FOR LOW SUPPLY VOLTAGE OP AMP APPLICATIONS

certain that the output current-handling capability of the gates is not exceeded. Either use high-value resistors or parallel a number of gates to increase the current capacity. For example, there are four *NOR* gates in the CD4001, so that three gates could be parallelled to increase output current drive (the parallel gates are at the very output, with the first gate a single since the current demand at this point is minimal).

In conclusion, the CMOS family of ICs is versatile and offers distinctive features. For a given application, consideration should be given to the exact system requirements to determine the family that best fits the requirements. There is no question of whether CMOS or TTL is better in an overall sense since a direct comparison clearly establishes the fact that each family offers distinct advantages and drawbacks. Basically, the strong points for CMOS are low power dissipation and wider noise margins. The drawbacks are evident in higher frequency operation where the low power advantage is lost and the highest frequency of operation is lower than that for TTL. With wider use of CMOS, the price is bound to decrease, causing the cost advantage enjoyed by TTL to all but disappear; this will establish CMOS as the undisputed choice in certain applications.

REVIEW QUESTIONS

5.1   What are the advantages of the CMOS logic family? What are the disadvantages? (Compare to the regular TTL.)

5.2   What is the typical threshold voltage for MOS transistors inside CMOS gates?

5.3   Sketch the transfer characteristics for a CMOS inverter with supply voltages of: (a) 5 V, (b) 10 V, and (c) 15 V.

5.4   Determine the power dissipation for a single CMOS inverter (type 74C04) as a function of the loading of other similar inverters. Let the fan-out vary between zero and 10. Graph the power dissipation vs. fan-out. Make three plots: at frequencies of 10 kHz, 100 kHz and 1 MHz. Make a conclusion from the plots. (Use a 5 V supply.)

5.5   Make a power dissipation comparison per inverter gate between CMOS (74C04) and standard TTL (7404). Assume both are operated from 5 V supplies and the maximum frequency is 5 MHz.

5.6   Why is it important to apply the supply voltage to CMOS circuits prior to applying an input? What may be the consequences of doing the reverse?

5.7   Compare the operation of inverters in CMOS (74C04 operating from a 15 V supply) and low-power Schottky TTL (74LS04). Assume that the highest frequency of operation to be considered is 10 MHz and that the fan-out in each case is 10 similar gates, i.e. CMOS loaded by 10 CMOS gates, TTL loaded by 10 TTL gates. If the overriding consideration is minimum power dissipation, is there a frequency above which the low-power Schottky TTL implementation is superior? If so, what is that frequency?

5.8   Although most CMOS products have protection built in, what precautions should be used in handling CMOS ICs?

5.9   In addition to basic logic functions (*OR, AND*, etc.), which are also available in other logic families, the CMOS family offers one other unique gate not available in other families. What is it and what are its characteristics?

5.10 Assume that a load resistor is connected to the output of a CMOS transmission gate from the supply. Treat the resulting gate as a two-input device, one input being the control voltage, the other V_I. Determine the logic truth table.

5.11 Repeat Question 5.10 above if the load resistor is connected to ground instead of the supply.

5.12 Besides being used to implement different logic functions (see Questions 5.10 and 5.11 above), transmission gates have other important applications. What are they?

5.13 Consider the CD4016 operated from a 15 V supply. Specify the range of voltages required at the logic (control) input. Also specify the range of analog input voltages that can be switched.

5.14 When two CD4016s are connected in parallel, what is the typical *on* resistance and *off* leakage current?

5.15 Compare the CD4016 with the CD4066 in performance characteristics.

5.16 Compare the implementation of an edge-triggered D flip flop in CMOS

(74C74) and TTL (7474). Make a count of the individual devices used (assuming roughly the same chip area required for bipolar and MOS transistors), and estimate which requires a larger chip.

5.17 How can CMOS gates be used in low-power *linear* applications? What are the advantages of doing so?

6

Interface ICs

In the majority of IC applications, more than one type or category of ICs is used, which requires some sort of interfacing. Typical interfaces are: from analog to TTL or CMOS digital; from TTL to CMOS and from CMOS to TTL; from a logic output to a remote location; and from either TTL or CMOS to a readout.

This section is devoted to the examination of the more typical interface problems and the ICs used to solve them.

6.1   COMPARATORS

The link between analog and digital signals is provided by the comparator circuits. As a typical circuit, let us use the LM311 comparator, whose schematic is shown in Fig. 6.1. Note the use of a linear high-gain differential amplifier at the input, with a digital output. Two differential stages are formed by $Q1, Q2$ and $Q3, Q4$. The signal is further amplified $Q8, Q9$, and $Q11$. (Active current sources—current mirrors—are used throughout.) The signal is then buffered by $Q12$ and applied to the output transistor, $Q15$. The maximum output sink current is established by limiting $Q16$ and $R13$. Note that the output is an open-collector and should be connected to a load that is referenced to $+V_{cc}$. As is the case with most linear amplifiers (OP AMPs), the input offset voltage effects can be balanced out—the scheme, which is almost identical to OP AMPs, is shown in Fig. 6.2. Another useful feature not present in OP AMPs is the capability to strobe the comparator, as shown in Fig. 6.3. If the strobing transistor is turned on (by

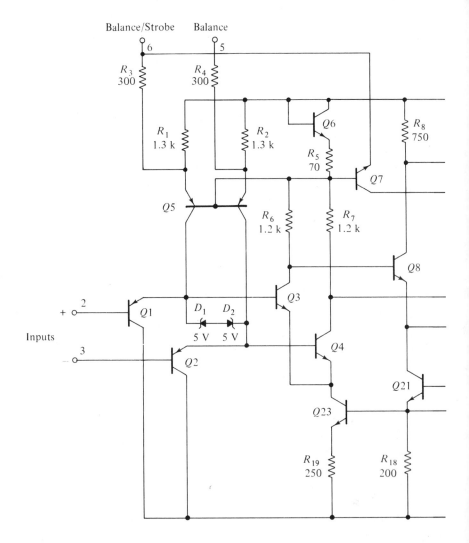

FIGURE 6.1   LM311 COMPARACTOR SCHEMATIC

driving its base high), pin 6, the balance/strobe input of the 311, is brought low. This has the effect of turning on $Q7$ which removes the current drive from $Q11$. Thus the current through $Q12$ and subsequently through the output transistor, $Q15$, is turned off. The output terminal under these conditions is unreferenced; that is, looking into the output terminal, we see a high resistance between the

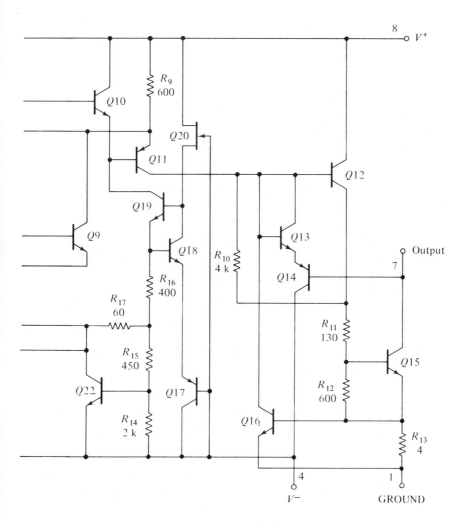

output and ground ($Q15$ off) as well as between the output and V_{cc}.

Typically, the comparator can be operated over a wide range of supply voltages: ± 15 V to a single $+5$ V supply (in this case $V-$ is connected to ground). The choice of power supply voltage is made on the basis of input and output requirements. If the output is to drive 5 V TTL, then a single 5 V supply

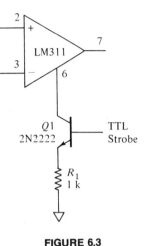

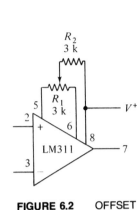

FIGURE 6.2   OFFSET
BALANCING

FIGURE 6.3
STROBING

would be used unless there were overriding considerations at the input: these would be the necessary range of input voltage. Neither input terminal should be more positive than $V+$, nor more negative than $V-$.

When driving digital gates, the output should be pulled up to $+V_{cc}$ with a resistor in the range of 1 k to 20 k. Note that for driving relays or LED indicators, the indicator is connected in series with the output and its current limiting resistor to $+V_{CC}$.

For a given application, the comparator is selected on the basis of all its operating parameters, but the primary ones are the response time and the voltage gain. The response time is the time delay between the application of the input condition and the time that the output attains the corresponding state. For the 311, the response time is typically 200 ns; comparators with response times in the order of 20 to 40 ns are also available (examples are the 710 and 306 comparators).

The comparator is intended to be used as a switching circuit, and as such, it has no frequency compensation, allowing for extremely short response times. It is therefore possible for the circuit to break into oscillation under conditions of very slowly varying inputs near the threshold voltage. Consider the non-inverting input to be biased at 2 V (with a single $+5$ V supply). If the inverting input is below 2 V, the output is high (assuming a pull-up resistor is used). If the inverting input is now brought above 2 V, the output switches low. With an output voltage swing of approximately 5 V (0 to 5) and a voltage gain of 200 V/mV, the amount by which the inverting input needs to exceed the threshold voltage (2 V in our example) is the output swing divided by the voltage gain, or 0.025 mV. However, should the inverting input signal remain near the 2 V threshold, the comparator is biased in its linear region, and oscillation will probably ensue. The 311 comparator is less susceptible to oscillation than comparators with faster response times. However, it is always a good idea either to precondition the signal so that it passes through the threshold quickly or to use positive feedback to provide hysteresis (see Schmitt trigger applications of OP AMPs).

To further reduce the possibility of oscillation without sacrificing response time, it is necessary to provide good isolation between input and output (by physically not running output leads near the input), and to bypass the power supplies (at the supply pins of the comparator) with low inductance capacitors (disc or solid tantalum). In addition, the lower the source resistance, the less chance for oscillation. Therefore, use as low a source resistance as the application will allow. This is in spite of the fact that comparators have relatively low input bias currents (typically 200 nA) and offer potentially very high input impedance. A wide-bandwidth high-source impedance system is inherently unstable; unnecessarily high source impedance should be avoided.

In addition to applications in detecting low-level signals and interfacing between logic families, comparators can be used in oscillator and multivibrator applications (see Section 2.12). While an OP AMP can be used as a comparator, its response times are too slow, typically in tens of μs, making the use of a comparator IC like the 311 mandatory.

6.2   LINE DRIVERS AND RECEIVERS

In many digital systems, one or more signals need to be transmitted from one location to another. The distance involved may be only a few feet or a few thousand feet; the basic problem involved in such transmission is to extract at the remote location the same signal as was sent, irrespective of the medium (type of line, i.e. twisted pair, coaxial) and the noise of the environment. Typical applications involve interfacing central processors with peripheral equipment.

Because transmission-line effects must be taken into account, the complete problem is beyond the scope of this section. Consider, then, the simplified picture shown in Fig. 6.4. The single-line system uses a TTL-type gate with a high-current open-collector driver transistor to drive the line. Typical of such transmitters are the 75450 and 75451 ICs. The receiver in this unbalanced system is a TTL inverter with an input pull-up resistor. Besides the transmission-line effects (the signal is delayed in time), there is attenuation of the signal in the line—this effect is represented by the line resistance R. The problem with this scheme is the drastic reduction in noise margin (for long lines), as well as the noise pickup in the line. (Electromagnetic noise is generated by electric motors, lights, TV and radio transmission, etc., so we can practically assume that noise signals are always present and at all frequencies.)

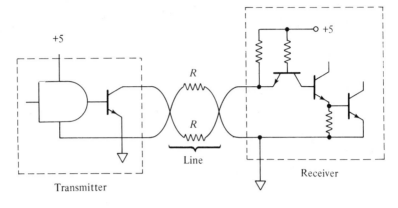

FIGURE 6.4   UNBALANCED TRANSMISSION

The balanced transceiver system shown in Fig. 6.5 overcomes some of the drawbacks of the unbalanced system. A three-line (two-line plus ground) transmission is required, with the signal developed by inversion. Again, the transmitter has TTL-type gates with open-collector high-current transistor drives capable of sinking 100 mA. The receiver is a differential amplifier, quite similar to a comparator. The line is terminated with R_T at the receiver. The basic advantage of this system is that any noise induced in the line is induced equally in the two leads, and appears as a common-mode signal at the receiver. The receiver is designed to have a high CMRR, thus allowing extraction of the desired signal from the line which has appreciable noise. In fact, signal-to-noise ratios of less

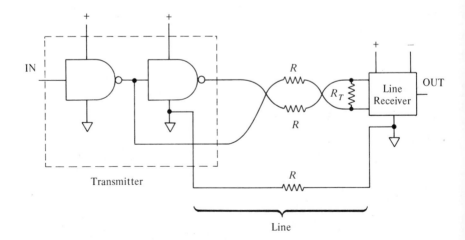

FIGURE 6.5   BALANCED TRANSMISSION

than one are not impossible to tolerate with the balanced system. Typical transmitters for the balanced system are 75110, and 75113; typical receivers are
75107 and 75115.

Of special interest is the *tri-state* or three-state type of driver and receiver.
To understand tri-state operation, first consider a TTL gate, the 74H00, for
example (see Fig. 4.8). With either input (or both) low, the drive for $Q2$ and $Q3$
is removed, and both $Q2$ and $Q3$ are off. The voltage on the base of $Q4$ is high;
therefore $Q4$ and $Q5$ are on, providing for a high at the output. Another way of
viewing this situation is to say that there is a very low resistance between the
supply and the output (when the output is in a high state). Similarly, with both
inputs high, drive is applied to $Q2$ and $Q3$, $Q4$ and $Q5$ are off, the output is low,
and $Q3$ is saturated. Another way of describing the same condition is to say that a
very low resistance to ground exists at the output. Thus the normal digital states
at an output may be considered in terms of the effective impedance to either the
supply or ground. The third possible state of the output, that provided in tri-state
gates, is that of a high resistance to *both* the supply and ground.

Consider the tri-state differential line receiver circuit shown in Fig. 6.6.
There are two distinct parts to the circuit: the front end made up of a differential
amplifier followed by emitter-followers to level-shift (together with the zeners)
and another differential amplifier with an output buffer. This is basically a
high-gain linear amplifier. The second part is a TTL Schottky-type gate; the
tri-state input (labelled strobe) is applied to a Schottky-TTL inverter and forms
one input of the *NAND* gate. As long as the strobe is low, the receiver functions
as a voltage comparator—the output is either high or low, determined by the
condition of the inputs. When the strobe is brought high, the output of the
inverter goes low. Without D_1 in the circuit, the output would be driven high.
However, D_1 conducts and clamps the base of $Q2$ to approximately 0.7 V. Thus
$Q2$ and $Q3$ are off due to the conduction of D_1; $Q1$ and $Q4$ are off due to normal
NAND action. The net result is a high impedance state at the output to either
ground or V_{cc}. Compare this to the operation of the line receiver shown in Fig.
6.7. The circuit is identical except that an open-collector *NAND* gate is used. The
strobe acts in much the same manner: if low, the inputs control the output (a
pull-up resistor is connected to V_{cc}); when the strobe is high, the inverter output
is low, and both $Q1$ and $Q2$ are off. Under these conditions, the output is in a
high resistance state (the third state).

The primary advantage of the tri-state receiver over the one in Fig. 6.7 is that
it has a conventional active pull-up, and therefore provides faster switching
(low-to-high) at the output when the receiver is enabled. (For a more detailed
discussion of the switching characteristics see Section 4.1.)

The usefulness of tri-state (or strobed) line drivers and receivers becomes
evident when bussed systems are considered. For example, we may use a single
line to transmit and receive signals: the line driver output and the receiver inputs
are connected in parallel at both ends of the line. When transmitting, the receiver

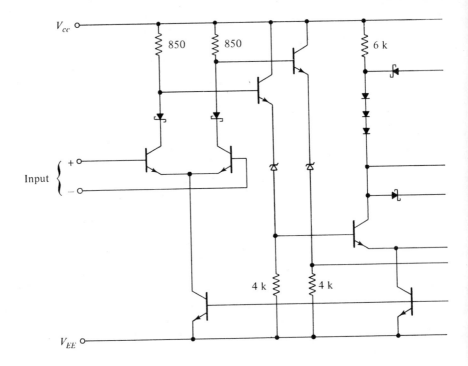

FIGURE 6.6   TRI-STATE DIFFERENTIAL LINE RECEIVER (DS3650)

is disabled and the driver enabled; when receiving, the driver is disabled and the
receiver enabled. Such systems are called party-line systems, since they share the
line. Another very common application involves connecting any number of tri-
state drivers with their outputs parallelled to a single line. Only one of the drivers
would be enabled at a time with its signal propagating to the receiver. An

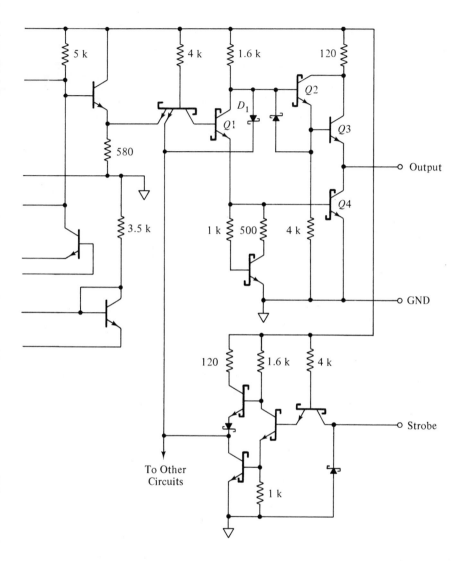

example of such an application is the data bus in a computer, where data from any one of a number of peripherals can be selected. Similarly, the data output from the computer may have a number of receivers in parallel across it, with each receiver located at a peripheral. By enabling the proper receiver, data from the processor is transmitted to the desired peripheral.

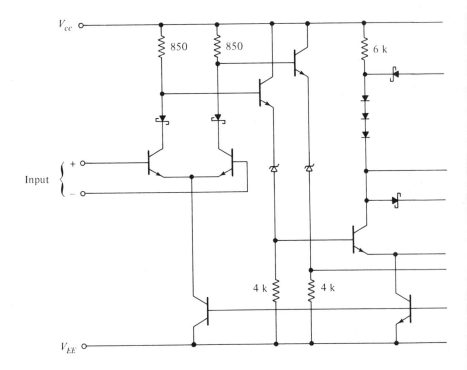

FIGURE 6.7   DIFFERENTIAL LINE RECEIVER (DS3652)

6.3   LOGIC TRANSLATORS

In many digital systems, a mix of logic families is used. TTL, CMOS and MOS all need to be interfaced to one another. The interfacing between TTL and CMOS can be direct if the CMOS is operated from the 5 V supply, since the

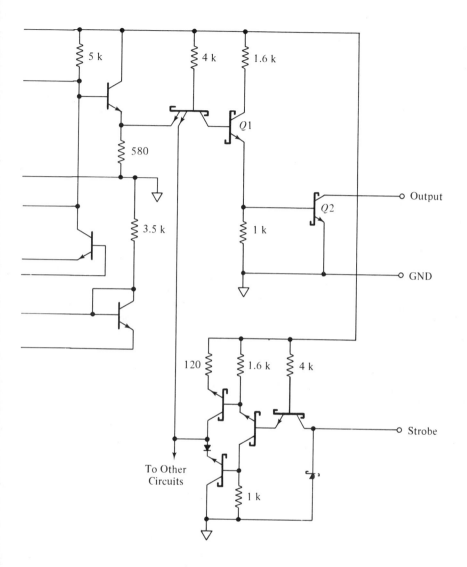

current demand by CMOS is negligibly small and the TTL output levels are compatible with CMOS input levels.

When interfacing CMOS to TTL, the output current capacity of CMOS usually is insufficient to drive standard TTL. Some CMOS gates do have sufficient current drive to interface directly to low-power TTL—examples are the CD4049 and CD4050 CMOS buffers. However, in general, most CMOS gates require some sort of current booster at the output in order to drive TTL. The DS

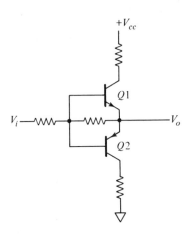

FIGURE 6.8   CMOS BUFFER ($^1/_6$ DS3630)

3630 buffer is designed for such applications. One-sixth of the circuit (the DS 3630 contains 6 buffers) is shown in Fig. 6.8. When the input is high, approximately 5 V from the output of a CMOS gate, $Q1$ conducts and provides an output approximately one diode drop lower than 5. When the input is low, typically close to zero, $Q2$ conducts and provides an output that is approximately one diode drop above ground. The logic levels at the output are compatible with TTL input levels. In addition, the output can sink 16 mA in the low state, thus providing a fan-out of 10 for TTL. The typical connection for the buffer interfacing CMOS to TTL is shown in Fig. 6.9.

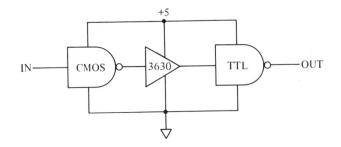

FIGURE 6.9   CMOS-TO-TTL INTERFACE

Another common interfacing problem is that of level-shifting from TTL (0 to +5 V) to *P*-MOS memories, shift registers, and switches, requiring negative voltage levels. Figure 6.10 shows the level translator circuit. The front end is TTL compatible: input diodes act much like the emitters of the TTL multiple emitter transistor, and the voltage at the point labelled X is established by three

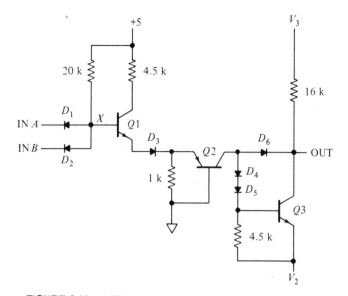

FIGURE 6.10   TTL-TO-MOS TRANSLATOR (½ DS8800)

diode drops to ground (the base-emitter drop of $Q1, D_3$, and the base-emitter drop of $Q2$). The level translation is accomplished in the *PNP* transistor $Q2$. If either input or both inputs are low, $Q1, D_3$, and $Q2$ are off. Therefore, D_4, D_5, and $Q3$ must also be off (D_6 keeps the V_3 supply from turning on $Q3$ under these input conditions). The output is at V_3 (minus the very small drop across the 16 k resistor).

When both inputs are high, $Q1, D_3$, and $Q2$ are all on. The collector current of $Q2$ turns on D_4, D_5, and $Q3$. As $Q3$ saturates, the output is forced to approximately 0.1 V above the V_2 supply. A wide latitude in the choice of supply voltages, V_2 and V_3, is tolerated by the circuit. The minimum differential is 5 V, the maximum 40 V. In addition, V_2 must always be more negative than V_3 (by 5 V), and in an absolute sense, V_2 must be more negative than -8 V. Once V_2 is established, the choice of the V_3 supply is determined by the minimum and maximum differential between V_2 and V_3.

6.4   LED INTERFACING

Along with developments in ICs has come a literal revolution in display technology. Individual light-emitting diodes (LEDs), seven-segment numeric LED readouts, as well as alphanumeric LED (array) readouts are available in various sizes and colors. Typical applications involve interfacing an LED display to a digital output. If the output is TTL, the high current capability makes the interfacing a simple problem: an open-collector gate may be used, with the LED and a current-limiting resistor connected in series to the 5 V supply. (In fact, a regular totem-pole output TTL gate can be used.) For seven-segment numeric readouts, the 7446, 7447, and 7448 BCD decoder-driver ICs are used. These ICs are TTL compatible at the input and accept a four-bit BCD (binary coded decimal) code, which is converted into a numeral (0 through 9) in the now standard seven-segment format. The 7446 and 7447 decoder-drivers are open-collector at the output, intended to drive common-anode LED readouts; the 7448 has a passive pull-up and can interface to common-cathode readouts.

The CMOS counterpart to these decoder-drivers is the CD4511. Its input is CMOS compatible; the IC accepts a 4-bit BCD input code and provides a 7-segment decoded output. Internal interfacing is accomplished by including an *NPN* transistor (on each output) to buffer the *P* MOSFET—thus direct interfacing (with a series current-limiting resistor) to common-cathode LED readouts is possible.

Another very common interfacing problem involves mating MOS output levels to LED readouts, where the multiple display data is in a decoded 7-segment format, but is multiplexed. Examples of this are clock, calculator, DVM (digital voltmeter), and other chips. To see the need for multiplexing, consider a clock application with a six-digit readout: if each digit were decoded inside the clock chip and brought out, 42 pins (6 times 7) would be necessary just for the output. In a multiplexed configuration, the seven-segment lines would be brought out, with six-digit lines, or a total of 13 lines as opposed to 42. In the multiplexed mode, the output data is provided as follows: only one of the digit-drive outputs is on at a given time, and during that time the seven-segment outputs contain the code for that particular digit. In the six-digit clock example, the period (established internally) is divided into six equal parts. During the first sixth of the period, the digit 1 output is enabled with the seven segment outputs corresponding to the proper code for the first digit. During the second sixth of the period, the digit 2 output is enabled, with the seven segment outputs now providing the code for the second digit. The cycle proceeds enabling the third, fourth, fifth, and sixth digits in turn. Once the period is completed, the cycle starts over again and repeats. Note that in such applications, each digit is on for only a fraction of the time—this fraction is called the *duty cycle*. In the clock example, the duty cycle for each digit is 1/6. (The strobing period is quite short, so that to the eye, all digits appear to be continuously illuminated although they are being strobed.) The light output of the LED readout is a function of the average power delivered to the LEDs. Since the voltage drop across a conducting LED is essentially constant (approximately 1.7 V), it is the current which establishes the light output. To obtain the same light output under multiplexed operation as under continuous operation, the current through the LED must be increased by a factor equal to the inverse of the duty cycle. This is accomplished by decreasing the series current-limiting resistor. In the clock example, if we wanted the same light output as obtained with a 10 mA continuous current through the LED readouts, we would have to decrease the series resistors to provide a current of 60 mA through each segment in order to provide the same average power.

There are a number of ICs especially designed for such applications. The 75491 quad segment driver, and 75492 hex (six) digit driver are but two examples. The respective circuits are shown in Figs. 6.11 and 6.12. The two circuits are almost identical—both utilize the current amplification of a Darlington connection. The 75491, when used as a segment driver (for a common-cathode readout), would have out 2 connected to $+V_{cc}$ and out 1 to the anode of the readout. (Two 75491s are required to drive the seven numeric segments with one driver left over; this last driver is used to drive the decimal point LED, if used.) Note that the 75491 can also be used as a segment driver without 1 connected to ground.

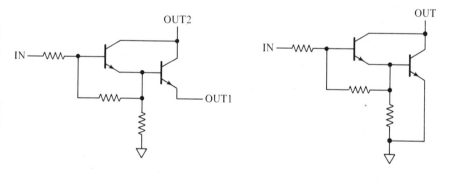

FIGURE 6.11   SCHEMATIC OF 75491
DRIVER (¼)

FIGURE 6.12   SCHEMATIC OF 75492
DRIVER (¹/₆)

A typical application is shown in Fig. 6.13. Each of the seven-segment outputs is applied to a 75491 driver (only one of these is shown). Each 75491 driver in turn is connected to one of seven LED anodes (plus the decimal point, if used). The common cathode of each readout (only one shown) is connected to the output of the 75492 driver; the input is derived from the MOS circuit. Each of n readouts is connected as follows: all the a anodes are common and driven by the a-segment output of the MOS circuit, after it has been buffered by 1/4 of a 75491. The b anodes of all the displays are similarly tied together and driven by 1/4 of a 75491. The same applies to all other anodes, i.e. the c-segments are all tied together, etc. To be on, a given segment LED must have both the segment drive high as well as the digit drive high. For example, the a-segment LED of the first readout (shown) will be on if the a-segment output of the MOS circuit is high *and* the digit 1 output is also high. This scheme can be extended to any

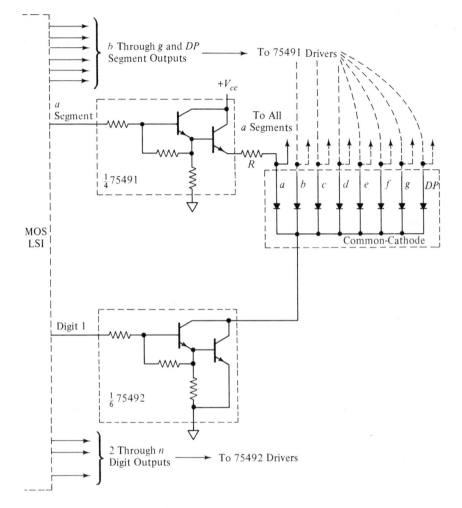

FIGURE 6.13   TYPICAL MULTIPLEXED READOUT APPLICATION

number of digits—two 75491s are needed for the seven segments plus decimal point, and a single 75492 can accommodate up to six digits—two 75492s up to 12 digits. Multiple LED digit displays are available with the anode inter-connections for multiplexed operations made internally.

The maximum source current for the 75491 is 50 mA; the maximum sink capability of the 75492 is 250 mA. Although the 75491 can be used as a digit driver for common-anode displays, the segment current limitation makes this application of the 75491 impractical. A discrete *PNP* transistor capable of sourc-ing 250 mA should be used as the digit driver; the 75492 can then be used as a segment driver.

6.5 OPTO-ISOLATORS

Opto-isolators (also called opto-couplers) are solid-state devices, incorporating a galium-arsenide emitter (LED) and a phototransistor sensor. Other sensors used are photodiodes, photodarlington transistors, and light-activated SCRs (the latter are used where significant current is switched). Opto-isolators are used to transmit digital signals while allowing electrical isolation between the input and output.

The 4N25 opto-isolator is typical, shown schematically in Fig. 6.14. The input is applied to the emitter diode (strictly speaking it is not an LED since its output is in the infrared region and not in the visible spectrum). If the output of the TTL gate in Fig. 6.14 goes low, the emitter diode is turned on and emits photons, which in turn are sensed by the transistor. In effect, the photons reaching the transistor act in much the same manner as conventional base current drive—the transistor turns on, saturates, and provides a low input to its TTL gate. Thus the digital signal has been transmitted through the opto-isolator without any electrical connection between the input and output. The ground or common at the input can be hundreds of volts higher or lower than the common reference at the output.

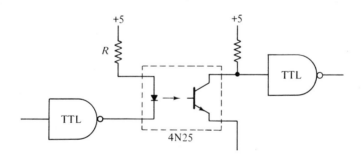

FIGURE 6.14   TYPICAL OPTO-ISOLATOR APPLICATION

The primary characteristics of opto-isolators are the propagation delay (switching speed) and the isolation voltage. Propagation delays of as little as 10 to 20 ns for photodiode detectors to 1 to 10 μs for phototransistor types like the 4N25 are available. The propagation delay and the current gain are directly related: the smaller the current gain (ratio of output current to the input emitter-diode current), the faster the switching (lower propagation delay). The 4N25 has a current gain of approximately 1; with an input forward current of 10 mA, the typical output collector current is 10 mA. (The current gain is totally analogous to the transistor β.)

The isolation voltage rating, 2500 V for the 4N25, is the maximum voltage difference between the input and output.

The opto-isolator is basically a solid-state replacement for a relay, and offers the advantages of no moving parts, and therefore more reliable operation, and faster switching speeds.

REVIEW QUESTIONS

6.1 What general function do interface ICs perform? Give some typical examples.

6.2 Describe the operation and uses of a comparator. In what ways are comparators similar to OP AMPs? In what way do they differ from OP AMPs?

6.3 If a comparator output is unstable or oscillates in a given application, what are some of the possible remedies?

6.4 An analog signal is to be interfaced to drive a TTL gate. The signal has typically a 50 mV (peak-to-peak) residual noise component. The desired indication is when the input exceeds (or falls below) approximately 2 V. Implement a comparator circuit to meet the requirements.

6.5 If digital signals have to be transmitted over any length of wire, what are the interface ICs that should be used and how?

6.6 Describe the operation of tri-state outputs. What are the uses for circuits possessing tri-state outputs?

6.7 State the similarities as well as the differences between strobed and tri-stated outputs. Which has better (faster) switching characteristics and why?

6.8 Which ICs are used in interfacing from one logic family to another?

6.9 Give examples of how CMOS gates can be interfaced to TTL as well as the other way around. Which of these would be more common and why?

6.10 What special problems need to be overcome in driving LED indicators?

6.11 A seven-segment LED readout group of four digits is to be multiplexed from a CMOS IC. The outputs are: seven lines for the segments a through g; four digit outputs, only one of which is on at a time. For desired light output each LED segment requires 20 mW of power (assume the voltage drop across the LED to be 1.7 V). Use 75491-2 interface ICs and provide a block diagram to implement the system (specify the limiting resistors, if used).

6.12 List the types of applications for opto-isolator ICs.

6.13 Specify the isolation rating, current gain and propagation delay for a 4N25 opto-isolator.

6.14 Show how an opto-isolator (like the 4N25) can be used to isolate two logic circuits, one TTL, the other CMOS. Give the circuit; specify resistor values.

7

Special Function
ICs

In this section, we shall examine a group of ICs which are representative of an ever-growing number of readily available chips which incorporate tremendous versatility and large-scale functions. With the availability of larger integrated functional blocks, the IC user becomes truly liberated in terms of the scope of project that he can reasonably attempt. The function that might otherwise require a few weeks or even months to implement with discrete components is available in a single package. Once basic familiarity with the operation of the "super" chips is attained, the only limitation in the applications and projects in which they can be utilized is one's imagination.

7.1   THE TIMER

The 555 IC, which may be numbered differently depending on the manufacturer (SN72555, LM555, etc.), is an extremely versatile timing building-block. It incorporates the functions shown in the block-diagram of Fig. 7.1 in an eight-pin dual-in-line package. The schematic is shown in Fig. 7.2. Comparator 1 is formed by transistors $Q1$ through $Q4$; comparator 2 by PNP transistors $Q7$ through $Q10$ (transistors $Q5$, $Q6$ and $Q12$, $Q13$ are active loads for the respective comparators). The FF is formed by transistors $Q17$ through $Q20$—the reset input is to the base of $Q17$, the set to the base of $Q18$. Transistor $Q23$ is a split-load phase inverter which provides the necessary signal splitting to drive the totem-pole output stage comprised of $Q24$, $Q26$, $Q27$, and $Q28$. Output current (with the output high) is provided by the Darlington connection of $Q27$ and $Q28$; with

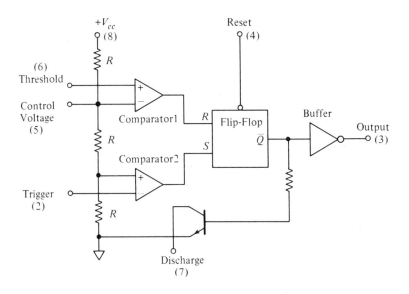

FIGURE 7.1   555 TIMER BLOCK DIAGRAM

the output low, output current is sunk by $Q24$ and $Q26$. Either sourcing or sinking, the maximum output current is 200 mA. External reset is provided by $Q25$. When the output is low, the discharge transistor, $Q14$, is on.

The comparator trip-points are established by the voltage divider of R_3, R_4, and R_5. Although these three resistors may not be exactly 5 k, they are fabricated at the same time and will be matched (probably to better than 2%).

The 555 may be operated in many different modes. Consider first the monostable configuration shown in Fig. 7.3. Timing is provided by the external RC combination; the undisturbed state is output low, Q high, the discharge transistor on, and the external capacitor at approximately ground. Pin 5, the modulation input, is bypassed to ground with a capacitor to prevent pick-up of stray signals. Triggering is accomplished by bringing the trigger input (pin 2) low. Normally, pin 2 is maintained high (near V_{cc}). When the trip-point for comparator 2 (see Fig. 7.1) is exceeded, that is, when pin 2 becomes more negative than $1/3\ V_{cc}$, the FF becomes set, Q goes low, the discharge transistor turns off, and the output goes high. At this point, the capacitor begins to charge through R toward V_{cc}. T, the duration of the output pulse, is determined by how long it takes the capacitor voltage to reach the trip-point of comparator 1. With an initial voltage of zero charging towards V_{cc} and a voltage of $2/3\ V_{cc}$ at time T, we have the following expression for the voltage across the capacitor:

$$\frac{2}{3}\ V_{cc} = V_{cc}\left(1 - e^{-T/RC}\right) + \frac{1}{3}V_{cc}$$

Note that V_{cc} drops out of the timing equation to give a pulse duration which is not a function of the supply voltage.

Solving for T, we obtain the pulse duration:

$$T = RC \ ln2 = 0.693 \, RC$$

A time T after the one-shot has been triggered, the capacitor voltage has reached $2/3 \, V_{cc}$, the FF is reset and the output is again low, with $\overline{Q}$ high. The discharge transistor turns on, and abruptly discharges the capacitor. The circuit has then returned to its original state.

The timing waveshapes for the 555 used as a one-shot are shown in Fig. 7.4. Note that the Darlington pair in the output pull-up circuit provides approximately 1.7 V reduction below V_{cc} in the output high-state voltage.

The trigger signal should be short, that is, pin 2 must be brought below $1/3 \, V_{cc}$ only long enough to cause the FF to be set (100 ns should be sufficient). The minimum reliable pulse width attainable is limited by the delays of the comparators, the FF, and the output buffer, and should be at least $10 \, \mu s$. The maximum pulse width is dictated by the quality of the capacitor, that is, the leakage resistance of the capacitor. As a general rule, the timing resistance should be smaller than the capacitor leakage resistance by a factor of at least 100. In addition, the timing resistor cannot be so large as to provide a charging current which is comparable to the threshold (or for the astable, the trigger) current, which is typically $0.5 \, \mu A$. Therefore, the maximum pulse width is practically limited to 10 to 100 s. For extremely large delays (pulse widths), the output can be extended by the use of counters. Such an arrangement is provided by the XR2240, which, in addition to having a timer similar to the 555, has an eight-bit counter on one chip.

For normal one-shot applications, the external reset line, pin 4, is maintained high, as shown in Fig. 7.3. However, it is possible to terminate the pulse at any time prior to T simply by strobing pin 4 low. This external reset adds flexibility and in some applications eliminates the need for additional logic circuits.

The basic astable connection of the 555 is shown in Fig. 7.5. Here both the trigger input and the threshold input are tied together to sense the capacitor voltage. Operation is as follows: capacitor charging occurs through both resistors R_1 and R_2; the capacitor is discharged through R_2. The lower limit of the capacitor voltage is $1/3 \, V_{cc}$, established by the trip-point of comparator 2; the upper limit of the capacitor voltage is $2/3 \, V_{cc}$, established by the trip-point of comparator 1. If we label the time the output is high as T_1 with a time constant equal to $(R_1 + R_2)C$ (capacitor charging) and label by T_2 the time the output is low (capacitor discharging) with time constant R_2C, the timing is determined by noting that the capacitor charges from $1/3 \, V_{cc}$ toward V_{cc}, and at T_1 its voltage is equal to $2/3 \, V_{cc}$. Thus:

$$\frac{2}{3} V_{cc} = (V_{cc} - \frac{1}{3} V_{cc})(1 - e^{-T_1/(R_1 + R_2)C}) + \frac{1}{3} V_{cc}$$

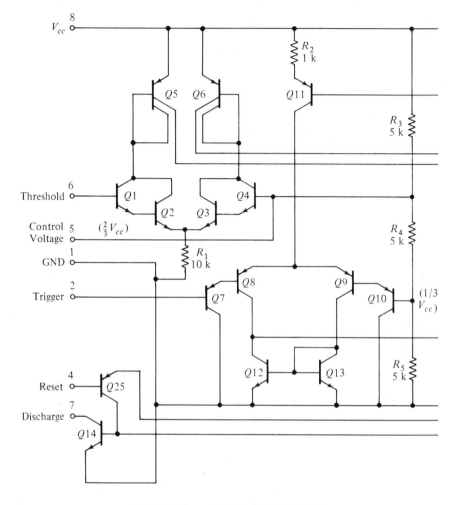

FIGURE 7.2   555 TIMER SCHEMATIC

Again, V_{cc} cancels, providing timing independent of the supply voltage:

$$T_1 = (R_1 + R_2)C \; ln2 = 0.693 \, (R_1 + R_2)C$$

Similarly, while the output is low, the capacitor discharges from $2/3 \, V_{cc}$ toward zero, and its voltage is equal to $1/3 \, V_{cc}$ at time T_2 later. The time constant is now R_2C; otherwise operation is symmetrical to the charging. The time T_1 is then:

$$T_2 = R_2C \; ln2 = 0.693 \, R_2C$$

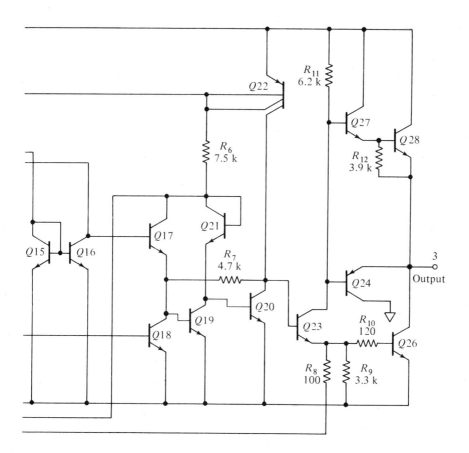

The period of the output square wave is then the sum of T_1 and T_2:

$$\text{Period} = 0.693 \ (R_1 + 2R_2)C = 1/\text{frequency}$$

The duty cycle is the ratio of time in the high state to the period, and is given by:

$$\text{Duty Cycle} = \frac{T_1}{T_1 + T_2} = \frac{R_1 + R_2}{R_1 + 2R_2}$$

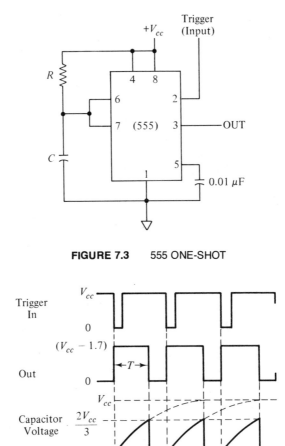

FIGURE 7.3   555 ONE-SHOT

FIGURE 7.4   ONE-SHOT WAVESHAPES

Note that although the duty cycle is easily adjustable by varying either R_1 or R_2, it is always less than 50%. It is obvious from the equation that for a 50% duty cycle, R_1 would have to be zero. This cannot be implemented, since when the discharge transistor (collector at pin 7) would turn on and try to saturate, it would be trying to pull the supply voltage to ground. This cannot occur because excessive current would flow into the discharge transistor, causing permanent damage to it.

The timing waveshapes for the astable configuration are shown in Fig. 7.6. Note that the output in the high state does not go all the way to V_{cc}, but typically is 1.7 V lower. This is still sufficient to drive TTL directly, assuming a 5 V supply is used.

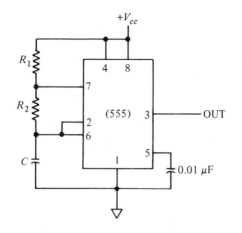

FIGURE 7.5   555 ASTABLE CONFIGURATION

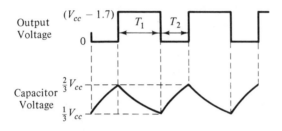

FIGURE 7.6   ASTABLE WAVESHAPES

The basic astable circuit must be modified if exactly 50% duty cycle is desired. A number of different schemes are possible; one is shown in Fig. 7.7. Here, charging is through R_1; discharging is through the parallel combination of R_1 and R_2. Since the equivalent voltage seen by the capacitor during discharge (pin 7 at ground) is $V_{cc}R_2/(R_1 + R_2)$, and since this voltage must be lower than $1/3\,V_{cc}$, R_1 must be greater than at least $2R_2$ or else the circuit will not oscillate. (If the equivalent voltage is not lower than $1/3\,V_{cc}$, the lower comparator trip-point is never reached and the output will latch in the low state.) The specific ratio between R_1 and R_2 for 50% duty cycle is calculated by equating the charging and discharging times. The resulting equation cannot be solved in closed form; iterative means yield the result that $R_1 = 2.362R_2$. In practice, this ratio is set by a potentiometer adjustment. The output frequency for this configuration is $0.72/R_1C$.

The circuits shown for astable operation have the external reset terminal tied to V_{cc}. It is possible to disable the square-wave output, that is, turn it off, simply by

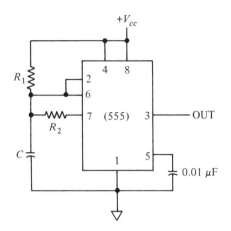

FIGURE 7.7   555 ASTABLE CONFIGURATION FOR 0.5 DUTY CYCLE

bringing the external reset line (pin 4) low. This causes the output to be low for as long as the reset line is maintained low. As soon as pin 4 is brought high, the output goes high and stays high for a time $1.1(R_1 + R_2)C$ for Fig. 7.5 and $1.1R_1C$ for Fig. 7.7; thereafter, normal timing resumes. (The assumption here is that the reset line is maintained low long enough for the capacitor to discharge fully; if that is not the case, the time in the high state just after the reset is made high is between the normal time and that specified above.)

This covers the basic operation of one-shot and astable configurations. In either application, the timing can be modified by using the modulation input, pin 5. The application of a voltage (from a low source impedance like the output of an OP AMP) to pin 5 causes the trip-point for both comparators to change, modifying the timing. When the modulation input (pin 5) is increased above its normal value of $2/3\ V_{cc}$, the timing is extended (lengthened); conversely, when the modulation input is made lower than $2/3\ V_{cc}$, the timing is shortened. In the astable mode, the modulation input acts to lengthen or shorten both T_1 and T_2. When the timing is modified through the modulation input, a well-regulated supply should be used, since the timing then is not independent of the supply. In addition, the relationship between the modulating voltage and the output pulse width or frequency is not a linear one. When not used, the modulation input should be bypassed to ground with a capacitor to prevent pickup of unwanted signals which would interfere with the proper timing of the circuit.

The 555 timer IC has a multitude of applications. The application notes available from the manufacturers contain many useful circuits. In addition, it is hoped that the understanding gained from the operating description above will allow the user an even wider range of applications through his own ingenuity and resourcefulness.

7.2   VOLTAGE-CONTROLLED OSCILLATOR

A voltage-controlled oscillator (VCO) is a device that provides an ac output signal, usually a square wave, whose frequency is directly and linearly proportional to the input voltage. The 555 timer, although it can provide an output frequency which is proportional to the modulation input, does not really qualify since the conversion from input voltage to output frequency is not linear, nor does it cover a wide range.

The 566 VCO IC, also available in an eight-pin DIP, provides linear voltage-to-frequency conversion. Its complete schematic is shown in Fig. 7.8; the block diagram is shown in Fig. 7.9. Two outputs are provided: a square wave (pin 3) and a triangle wave (pin 4). Basic operation is established by connecting an external capacitor from pin 7 to ground and an external resistor from pin 6 to V_{cc}, and by applying the input voltage to the modulation input, pin 5. The current source, formed by transistors $Q3$ through $Q13$, provides equal charging and discharging currents to the capacitor. The magnitude of these currents is set by the timing resistor and the modulation input voltage as follows: the voltage at pin 6 is the same as that applied to pin 5 since the base-emitter drops of the NPN ($Q3$) and PNP ($Q6$) transistors cancel. Thus the voltage drop across the timing resistor is $V_{cc}-V_5$ where V_5 is the modulation input voltage.

The voltage across the capacitor is buffered by $Q16$ and applied to the Schmitt-trigger circuit formed by $Q17$ through $Q22$, as well as to an output buffer $Q1$. The Schmitt trigger is actually formed by $Q19$ and $Q22$, with $Q17$, $Q18$, $Q20$, and $Q21$ being used as diodes to prevent $Q19$ and $Q22$ from saturating, thus improving the switching speed of the Schmitt trigger. For example, if the emitter of $Q16$ is high, diodes $Q18$ and $Q21$ are on with $Q19$ on and $Q22$ off. Thus the collector-emitter voltage of $Q19$ cannot be less than one diode drop ($Q21$), and therefore $Q19$ cannot saturate.

The trip-points for the Schmitt trigger are established by resistors R_6, R_8, and R_{16}. When the capacitor is charging: $Q19$ is off, $Q22$ on; $Q23$ is off, $Q26$ on; $Q27$ and $Q29$ off. Therefore, the diodes ($Q14$ and $Q15$), as well as $Q13$, are all off. The current from the timing resistor passes through $Q6$, $Q7$, and $Q9$ to the capacitor (pin 7). As the voltage across the capacitor increases linearly with time (a constant current gives a linear ramp), the base of $Q19$ is increasing in voltage until it is high enough to cause $Q19$ to conduct. This point occurs when the capacitor reaches 3 base-emitter drops above the voltage on R_8. The voltage across R_8 with $Q22$ on is the supply minus the base emitter drops of $Q21$ and $Q22$, divided between R_8 and the parallel combination of R_6 and R_{16}. (The voltages at the two ends of R_6 and R_{16} are approximately the same since $Q20$ through $Q22$ are on.) For the values shown (which are only typical), the highest voltage attained by the capacitor, when $Q19$ has just turned on and $Q22$ has just turned off, is $0.5\,V_{cc}$. At this time, the

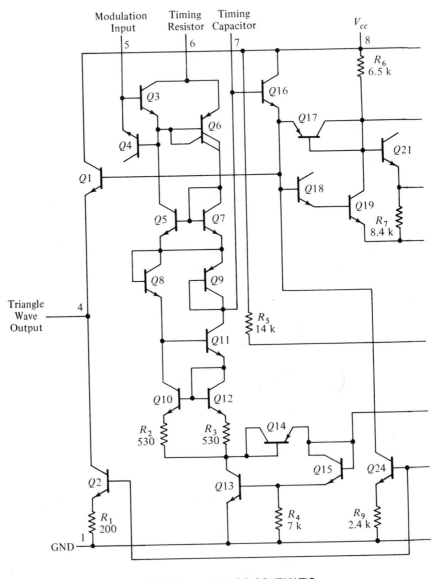

FIGURE 7.8   566 VCO SCHEMATIC

triangle-wave output voltage is two diode drops lower (due to $Q16$ and $Q1$).

Once $Q19$ turns on, $Q20$, $Q21$, and $Q22$ turn off. This in turn activates $Q23$ to conduct, which turns off $Q26$. Once $Q26$ is off, transistors $Q27$ and $Q29$, diodes $Q14$ and $Q15$, and transistor $Q13$ all conduct. This provides a path to ground for the timing current from the external resistor through $Q3$, $Q5$, $Q8$, and $Q10$, with $Q6$, $Q7$, and $Q9$ essentially off. The current mirror $Q10$, $Q12$ reflects the timing

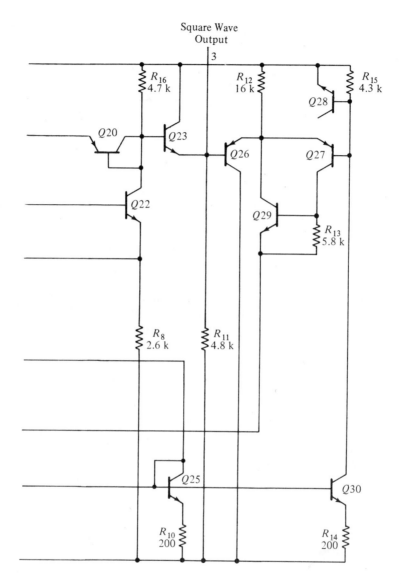

Square Wave
Output

current which passes through $Q10$ to $Q12$. The current through $Q12$ passes through $Q11$ and is drawn from the capacitor, thus discharging the capacitor. Note that due to the current-mirror action, the magnitude of the discharging current is the same as that of the charging current, and therefore is a constant proportional to the difference between the supply voltage and V_5.

As the capacitor discharges, the voltage on the emitter of $Q16$ decreases until

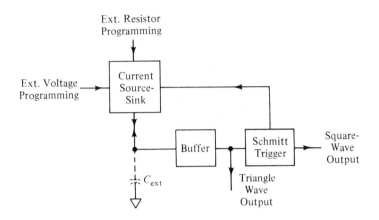

Ext. Resistor
Programming

Ext. Voltage
Programming

Current
Source-
Sink

Buffer

Schmitt
Trigger

Square-
Wave
Output

C_{ext}

Triangle
Wave
Output

FIGURE 7.9   556 VCO BLOCK DIAGRAM

$Q22$ turns on. At the same time, the triangle-wave output is two diode drops below this value. Since the charging and discharging currents are the same, the triangle-wave output and the square-wave output have exactly 50% duty cycles. With a constant current of $(V_{cc} - V_5)/R$, the time it takes to go from one threshold to the other is:

$$t = \Delta V \frac{C}{I} = (0.25V_{cc}) \frac{RC}{(V_{cc} - V_5)}$$

The period is twice this time, and the frequency is the inverse of the period, so we get the following approximate expression for the frequency:

$$f \cong \frac{2\,(1 - V_5/V_{cc})}{(1)\,RC}$$

The above expression is approximate since the values of $R_6, R_8,$ and R_{16} cannot be controlled exactly in manufacture, and the conversion of voltage (V_5) to frequency constant cannot be predicted exactly. However, from the above equation it is evident that the output frequency is linearly related to the input voltage. The approximate voltage-to-frequency conversion constant is:

$$V_i \cong V_{cc} - V_5 = f \frac{RC(V_{cc})}{2}$$

Pin 5 has a usable voltage range from just slightly less than the supply voltage to 3/4 of the supply voltage. The minimum is set by the voltage required to keep the current-source transistors biased. A typical application circuit is shown in Fig. 7.10; the corresponding waveshapes are shown in Fig. 7.11. Note that the square-wave output, obtained at the emitter of $Q23$, has limits of approximately $0.5V_{cc} - 2V_{BE}$ in the low state, and $V_{cc} - V_{BE}$ in the high state.

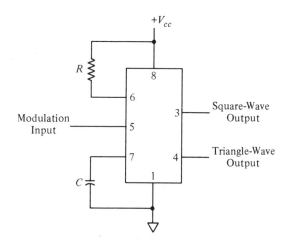

FIGURE 7.10   566 VCO (DIRECTLY COUPLED INPUT)

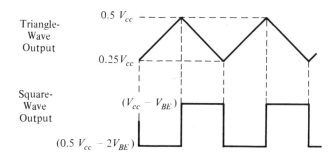

FIGURE 7.11   TIMING WAVESHAPES FOR 566 VCO

An alternate scheme for applying the modulating signal is to capacitively couple it to pin 5, having established a dc bias on pin 5, as shown in Fig. 7.12. For maximum input signal swing, positive and negative, the resistor ratio of R_1 and R_2 should be such that the voltage applied to pin 5 is $7/8\ V_{cc}$, or $R_2 = 7R_1$. In this manner, the net change in output frequency attainable is a factor of 10, that is, the highest frequency is 10 times the lowest. With $V_{cc} = 12$ V, the output frequency is given by:

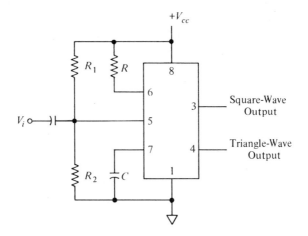

FIGURE 7.12   AC COUPLED MODULATION INPUT FOR 566 VCO

$$\frac{f}{f_o} \cong 1 - 0.667V_i$$

where f_o is the output frequency when the input is zero, and the range of V_i is plus or minus a maximum of slightly less than 1/8 of V_{cc}, in this case less than 1.5 V peak.

7.3   FUNCTION GENERATOR ICs

As defined here, a function generator provides square-, triangle-, and sine-wave outputs, which can be swept in frequency over a range of a number of decades.

The Intersil 8038 function generator chip, whose schematic is shown in Fig. 7.13, provides these waveshapes, operates at frequencies of 0.001 Hz to 1 MHz, can be swept over three decades, and has an adjustable duty cycle (from 2 to 98%). It is an extremely versatile building block and has almost unlimited applications.

The block diagram shown in Fig. 7.14 can aid in the understanding of the operation. Note that the circuit is similar to the 555 timer as well as the 566 VCO. The use of two comparators whose trip-points are set by three resistors R_8, R_9, and R_{10}, all 5 k, is the basic scheme used in the 555. The programmable current sources are in some ways similar to those used in the 566. External resistors between pins 4 and 5 and the supply set the current together with the programming voltage applied to pin 8. Note that the voltage applied to pin 8 is reflected by the *PNP* transistors ($Q2$ and $Q3$) to pins 4 and 5; thus the current through R_A and R_B is $(V_{cc} - V_8)/R_A$ and $(V_{cc} - V_8)/R_B$, respectively. If these two timing resistors are equal, then the

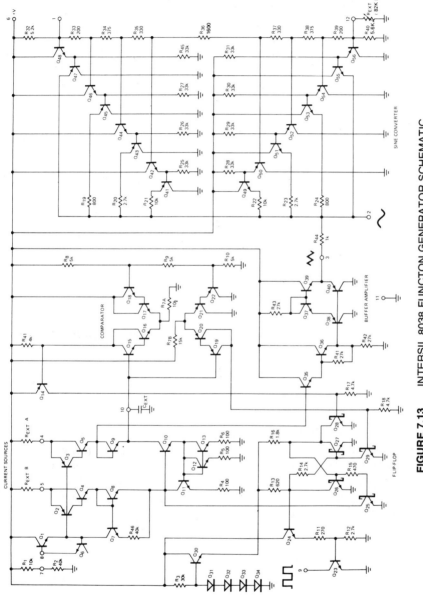

FIGURE 7.13   INTERSIL 8038 FUNCTON GENERATOR SCHEMATIC

207

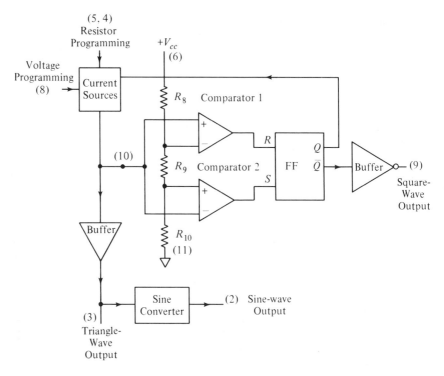

FIGURE 7.14   BLOCK DIAGRAM OF 8038 FUNCTION GENERATOR

currents are equal. While the capacitor at pin 10 is charging, $Q25$ is on (output at pin 9 high), and $Q10$ through $Q13$ current sources (actually sinks) are all off. Thus the charging current is that provided by $Q3$, $Q5$, and $Q9$—$(V_{cc} - V_8)/R_A$. When the capacitor reaches the threshold of comparator 1, $2/3\, V_{cc}$, the FF is reset causing $Q25$ to turn off. Now the current mirror, $Q10$ through $Q13$, is on. The current through $Q11$ is $(V_{cc} - V_8)/R_B$; transistors $Q12$ and $Q13$ mirror exactly twice this current to $Q10$. The conditions now are: current through $Q3$, $Q5$, and $Q9$ is still $(V_{cc} - V_8)/R_A$; current through $Q10$ is $2(V_{cc} - V_8)/R_B$. Therefore the capacitor is being discharged by a constant current equal to the difference between the currents through $Q9$ and $Q10$. The discharge current is:

$$I_{\text{discharge}} = (V_{cc} - V_8)\left(\frac{2R_A - R_B}{R_A R_B}\right)$$

Obviously, with $R_A = R_B$, the discharge current is equal to the charging current. The capacitor discharges until its voltage reaches the threshold voltage for comparator 2, $1/3\, V_{cc}$. At this point the FF is set, $Q25$ turns on, and the capacitor once again charges toward $2/3\, V_{cc}$. The timing during charging is determined by how long it takes the capacitor to charge from 1/3 to 2/3 V_{cc}, while charging with constant current $(V_{cc} - V_8)/R_A$. If we label this time by T_1, then:

$$T_1 = \Delta V \frac{C}{I} = \frac{V_{cc}}{3}\left(\frac{R_A C}{V_{cc} - V_8}\right)$$

In a similar manner we can determine the discharge time, T_2:

$$T_2 = \frac{V_{cc}}{3}\left[\frac{(2R_A - R_B)C}{V_{cc} - V_8}\right]$$

For 50% duty cycle, T_1 is equal to T_2; this is accomplished by adjusting $R_A = R_B$. For non-swept (constant frequency) applications, the voltage at pin 8 is supplied by the internal voltage divider made up of R_1 and R_2. Then $V_8 = 0.8V_{cc}$, and the above equations simplify to:

$$T_1 = \frac{5}{3} R_A C \qquad \text{and} \qquad T_2 = \frac{5}{3}(2R_A - R_B)C$$

Note that in all cases the timing is independent of the supply voltage. The frequency is calculated from the inverse of the sum of T_1 and T_2.

The square-wave output is derived from the FF, which is formed by the cross-coupling of $Q26$ and $Q27$. Transistors $Q28$ and $Q29$ provide for reset and set, respectively. The inverting buffer for the square-wave output is formed by $Q24$ and $Q23$. Note that $Q23$ provides an open-collector output, and as such should be pulled up with a resistor to the 8038 supply or any voltage within the range of the chip (less than 30 V); it need not be returned to the same supply as the 8038. Thus, for example, the 8038 could be operated from ± 12 V supplies and pin 9 could be pulled up to $+5$ V, making the square-wave output TTL compatible.

The triangle wave is derived from the capacitor voltage: the capacitor voltage is picked off one diode drop higher at the collector of $Q9$ and buffered by the class B amplifier consisting of $Q35$ through $Q40$. Note that the voltage at the triangle-wave output, pin 3, is the same as that across the capacitor: the base-to-emitter drops of $Q9$, $Q35$, $Q36$, and $Q40$ cancel. The output at pin 3 is a low impedance point—$Q39$ sources current and $Q40$ sinks it when necessary.

The sine converter, consisting of $Q41$ through $Q56$ and the associated resistors, is a complementary voltage-variable attenuator. Basically, to convert a triangle wave into a sine wave, the sharp peaks of the triangle wave must be distorted or rounded off. To accomplish this, the sine converter network provides an effective resistance between R_{44} and ground which is a function of the voltage at R_{44}. The voltage divider consisting of R_{32} through R_{40} sets the threshold voltages at different transistors which conduct and switch in additional resistance. The thresholds are symmetrical about $V_{cc}/2$, as are the two halves of the converter: transistors $Q41$ through $Q48$ act on the upper portion of the triangle and transistors

$Q49$ through $Q56$ act on the lower portion. To see the action of this attenuator, consider the triangle wave to be near $V_{cc}/2$ and increasing toward 2/3 V_{cc}. The voltage on the base of $Q42$ as well as the emitter of $Q41$ is at approximately $0.56V_{cc}$ (set by the voltage divider $R_{32} - R_{40}$). Thus when the triangle wave exceeds $0.56V_{cc}$, $Q21$ switches in R_{21}, providing for attenuation $R_{21}/(R_{44} + R_{21})$ to pin 2, the sine-wave output. As the triangle-wave voltage increases further, the attenuation increases with the switching in of additional resistors—R_{20}, etc. Thus there is more attenuation as the sharp peak of the triangle is reached, providing for the necessary rounding of the wave. Action is symmetrical for triangle inputs below $V_{cc}/2$, thus providing a symmetrically rounded waveshape at pin 2. Typically, the distortion of the sine-wave output is 1%; it can be reduced further by trimming the resistance between pin 12 and ground and between pin 1 and the supply.

Since the triangle and sine outputs are symmetrical about $V_{cc}/2$, the triangle amplitude $V_{cc}/3$ (peak-to-peak) and the sine amplitude approximately $0.22V_{cc}$, these two outputs are then centered about ground by using dual tracking supplies.

The timing waveshapes for the three possible outputs are shown in Fig. 7.15, where the adjustment has been made for 50% duty cycle.

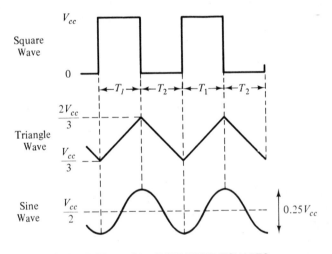

FIGURE 7.15   8038 OUTPUT WAVESHAPES

Figure 7.16 shows a number of connections for the timing resistors: in (a), two separate potentiometers can be trimmed to give the desired symmetry and output frequency; the configuration in (b) allows trimming for symmetry (50% duty cycle) while maintaining essentially constant frequency; in (c), a single resistor is trimmed to provide the desired frequency (no symmetry adjustment).

A linear voltage-controlled oscillator application is depicted in Fig. 7.17. A

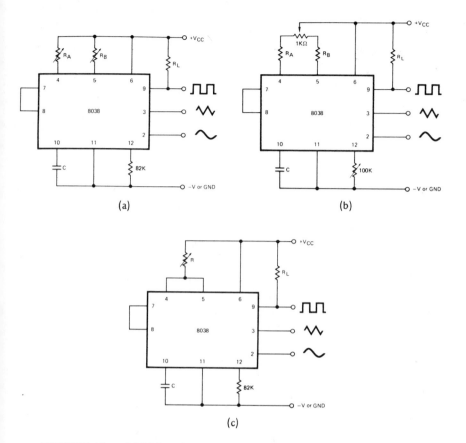

FIGURE 7.16   POSSIBLE CONNECTIONS FOR THE EXTERNAL TIMING
RESISTORS

741 OP AMP is used to buffer the input and sine output. The first 741 is used as a
voltage follower (the voltage at pins 8 and 4 should be the same); additional
compensation to insure stability is provided by the 1000 pF capacitor and any dc
offset is nulled by the 10 k potentiometer. For the configuration shown, the input
(sweep voltage) must be between zero (ground, giving the lowest frequency) and
−3 V (giving the highest frequency). Adjustments P_1 and P_3 set the symmetry at
the high and low frequency ends respectively, while P_2 trims the sine converter to
minimize harmonic distortion of the sine output, which is then ac coupled to the
741 output buffer. The ac coupling references the sine output to ground. (The
average level of the signal at pin 2 is −7.5 V; at the output it is 0 V.) The
triangle-wave output is 5 V peak-to-peak between −5 and −10 V; the square-
wave output, pulled up by the 10 k resistor, is at −15 V when low, and 0 V when
high.

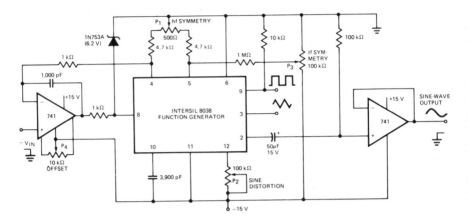

FIGURE 7.17   LINEAR VOLTAGE-CONTROLLED OSCILLATOR

In all applications requiring square-wave outputs in addition to triangle- and sine-wave outputs, the supply (or supplies) should be bypassed at the pins of the 8038 (pin 6) to minimize the spiking effect of the turn-on and turn-off of $Q23$. If the square wave is not used, the pull-up resistor from the square-wave output should be left out.

Quite similar in performance to the 8038 is the Exar 2206, whose schematic and block diagrams are shown in Figs. 7.18 and 7.19. Although the two ICs perform similarly from an external standpoint, the manner in which the functions are implemented internally reflect a different design philosophy. The heart of the XR 2206 function generator is the oscillator consisting of $Q1$ through $Q8$. Operation is as follows: pin 10 is at V_1 (typically regulated at 3 V) since the V_{BE} of the NPN ($Q9$) and PNP ($Q10$) transistors cancel. Similarly, the emitters of $Q2, Q3$ and $Q6, Q7$ are also at V_1. The two sides are symmetrical, i.e. $Q1$ through $Q4$ and $Q5$ through $Q8$. Any small amount of imbalance starts oscillation which alternately charges the external capacitor through $Q4$ and $Q8$. The FSK (frequency-shift keying) input at pin 9 determines whether the timing current is supplied through R_1 or R_2: if pin 9 is high, $Q11$ is off, $Q12$ on, $Q13$ off, $Q14$ off, and $Q15$ on. Since $Q13$ is off, the voltage at pin 7 is V_1 and $Q3$ or $Q7$ sink current to ground through R_1—this current is V_1/R_1. Since $Q15$ is on, the voltage at pin 8 is higher than V so that transistors $Q2$ and $Q6$ are both off. If a low input is applied to the FSK pin, the situation reverses—$Q3$ and $Q7$ are both off, while either $Q2$ or $Q6$ sink current to ground through R_2 (the current is V_1/R_2).

As soon as there is any charge on the capacitor, one side must be on while the other must be off. Consider $Q1$ on: current I then is drawn through R on the left side so that $Q4$ is off ($Q5$ and $Q6$ are also off with $Q3$ on and $Q2$ off—FSK input is assumed high for the succeeding discussion). The capacitor current is V_1/R_1 and flows through $Q8$, out of pin 5, through the external capacitor, into pin 6, and through $Q3$ and R_1 to ground. Since V_1 is fixed, the capacitor charges linearly—the voltage at pin 5 increases linearly with time, while the voltage at pin 6 decreases linerally with time. When pin 6 drops sufficiently (below $V_R - 3V_{BE}-IR$), $Q4$ begins to conduct. Immediately, $Q8$ turns off and $Q7$ turns on with $Q3$ turning off. The capacitor charging current now is through $Q4$, out pin 6, through the capacitor, into pin 5, through $Q7$ and R_1 to ground. Note that the current is the same as before: V_1/R_1 and constant. Thus, the capacitor voltage changes linearly again, with pin 6 increasing and pin 5 decreasing in voltage.

The situation persists until the voltage at pin 5 drops below $V_R - 3V_{BE} - IR$ and $Q8$ once again turns on. The net change in the capacitor voltage is then $2IR$. The current source I and the resistor R are designed so that $2IR=V_1/2$ or approximately 1.5 V. The half-period is then the time it takes to change the charge on the capacitor by $V_1/2$ charging at a constant current of V_1/R_1:

$$\frac{T}{2} =\left(\frac{V_1}{2}\right)\left(\frac{R_1}{V_1}\right)C = \frac{R_1C}{2}$$

Thus:

$$f = \frac{1}{T} = R_1C$$

By bringing the FSK input low, the output frequency is determined in much the same manner and depends on R_2 instead of R_1. The timing waveshapes are shown in Fig. 7.20.

Square-wave output is derived by the comparator connected to the bases of $Q1$ and $Q5$, and available externally at pin 11. Note that this is an open-collector output and needs to be pulled up with a resistor to a voltage more positive than that applied to pin 12 (which need not be grounded).

The triangle wave across the capacitor is taken differentially and amplified by two differential amplifiers consisting of $Q16$ through $Q19$. If no connection is made between pins 13 and 14, the amplification is linear and a triangle wave eventually emerges at pin 2. However, by connecting a low-valued resistor ($200\,\Omega$ typically), the amplification is not linear: the differential amplifier of $Q18, Q19$ is overdriven, and the sharp peaks of the triangle wave are rounded to produce a sine wave. Functionally, the overdriven amplifier in the XR 2206 performs the same triangle-to-sine conversion as the programmable attenuator in the 8038 discussed earlier.

The signal, either a sine or a triangle wave, is then fed through an analog multiplier consisting of $Q20$ through $Q23$ and its associated current sources. If pin 1, the other input of the multiplier, is grounded, the signal is unmodified by the multiplier. It is next applied to the differential output buffer amplifier consisting of $Q24$ through $Q28$. The output amplitude can be controlled by the dc potential applied at pin 3—nominally this should be at approximately half the supply voltage (or at ground if + and − supplies are used). In the case of a sine-wave output, the

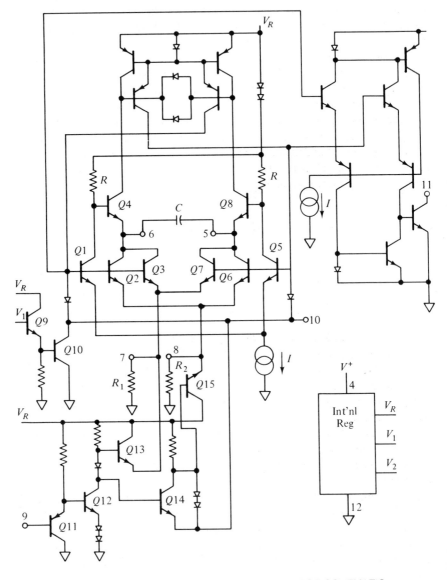

FIGURE 7.18   EXAR 2206 FUNCTION GENERATOR SCHEMATIC

distortion of the sine wave can be minimized by trimming the voltage at pin 3, as
well as by adjusting the offset voltage of the $Q16, Q17$ differential amplifier with a
potentiometer between pins 15 and 16 and the wiper connected to ground (or $-V_{cc}$
in dual supply applications).

A finer control of the output amplitude can be exercised by raising the voltage

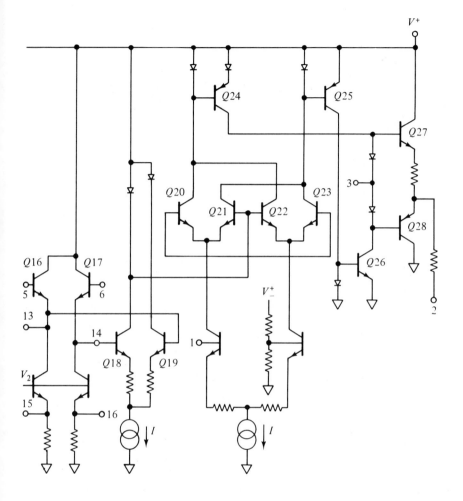

at pin 1 above ground. The waveshape, sine or triangle, is one input to the multiplier with pin 1 providing the other input. The output is then proportional to the product of the two inputs. One of the two ways in which this feature can be utilized is by trimming the dc voltage at pin 1 to provide the desired output amplitude. In this case, the dc voltage at pin 1, once adjusted, is not changed and

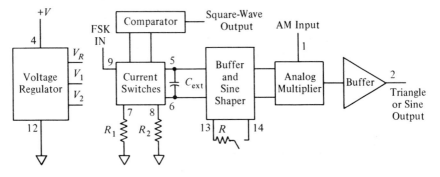

FIGURE 7.19   EXAR 2206 FUNCTION GENERATOR BLOCK DIAGRAM

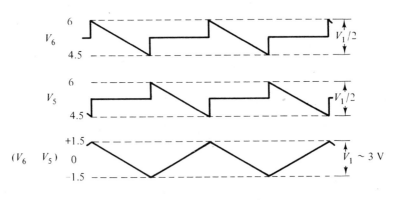

FIGURE 7.20   XR 2206 TIMING WAVESHAPES

should be bypassed to ground with 1 μF or higher capacitance. An even more useful application involving the analog multiplier is that of amplitude modulating the sine- or triangle-wave output. The relative sine- or triangle-wave output amplitude as a function of the voltage applied to pin 1 is shown in Fig. 7.21. (A 10 V supply is assumed.) To amplitude modulate the sine or triangle wave 100%, the modulating signal at pin 1 should have a peak-to-peak amplitude of approximately 0.4 of the supply voltage, and an average level at 0.3 of the supply voltage. Note that only the sine or triangle outputs can be amplitude modulated, and the square-wave amplitude is unaffected by the modulation input at pin 1.

The frequency of the sine or triangle wave, as well as that of the square wave, can be swept, i.e. all outputs can be frequency modulated simultaneously. The circuit configuration is shown in Fig. 7.22. The frequency is established by the total timing current I_T (3 mA maximum, 1 μA minimum), providing a maximum sweep range of 3000. When the sweep input V_C is zero, the output frequency is determined by the external capacitor value and the parallel combination of R and R_C—this is the highest frequency. When $V_C = V_1$ (approximately 3 V), $I_C = 0$,

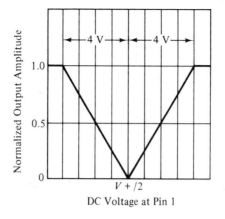

V + /2

DC Voltage at Pin 1

FIGURE 7.21   NORMALIZED OUTPUT AMPLITUDE VS DC BIAS AT AM
INPUT (PIN 1)

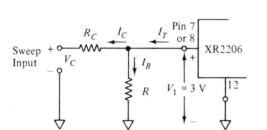

FIGURE 7.22   XR 2206 CIRCUIT CONNECTION FOR SWEPT-FREQUENCY
APPLICATIONS

and the timing is determined only by R, this is the lowest frequency. The functional
dependence of the output frequency on the sweep voltage is:

$$f = f_o \left[1 + \frac{R}{R_C} \left(1 - \frac{V_C}{V_1} \right) \right]$$

where $f_o = 1/RC$. If the sweep voltage is between 0 and V_1, then the sweep range is
adjusted simply by providing the appropriate ratio between R and R_C. For example, if $R/R_C = 9$, the sweep range is 10, or between f_o and $10f_o$. A simple sawtooth
generator suitable for use as the sweep voltage generator is shown in Fig. 7.23. OP
AMP 1 is used to integrate a negative voltage—the values of R, C, and $-V$
determine the sweep rate. Amplifier 2 is used as a comparator: as the output of A_1
ramps positive and reaches V_1, the comparator A_2 output goes high, turning on $Q1$.
The capacitor is discharged rapidly to zero, the comparator output goes negative,
and $Q1$ turns off, allowing the integrator once again to ramp from zero to V_1. (The

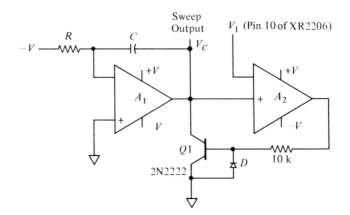

FIGURE 7.23   SIMPLE SWEEP GENERATOR FOR USE WITH XR 2206

diode protects $Q1$ while the comparator output is negative.) Almost any OP AMP is suitable; the 741 can be used if the sweep rate is not faster than a few ms.

The FSK input can be used to remotely program the function generator to one of two predetermined frequencies—the input is TTL compatible. If desired, the FSK input can be manually switched between the two frequencies. Another use of the FSK input is to generate square waves with other than a 0.5 duty cycle. For such applications, the FSK input is connected directly to the square-wave output (pin 9 to pin 11), remembering to use a pull-up resistor. In this manner, while the output is high, timing is determined by the resistor connected to pin 7; the time in the low state is determined by the timing resistor connected to pin 8. In this manner, the duty cycle is adjustable between 1 and 99%. With a high duty cycle, the triangle-wave output approaches a sawtooth waveshape.

To summarize, the two monolithic function generators discussed are representative of the circuit complexity and capability that is available at a fraction of the cost and size of the discrete-component versions. The emphasis here has been on understanding internal operation to enhance and enable the use of these ICs in a multitude of applications.

7.4   DVM ICs

Most IC manufacturers produce A/D (analog to digital) converter ICs, and some also have DVM (digital voltmeter) chips. We shall discuss some of the aspects of two such DVM chips: the Siliconix LD130 and the National Semiconductor MM5330–LF11300. The importance of DVM chips cannot be overemphasized: they have made the analog meter movement almost obsolete since in most applications they offer superior performance and reliability (no moving parts) at only slightly higher cost.

The most popular A/D scheme in voltmeter applications is the dual-slope conversion. Although relatively easy to describe, this scheme requires numerous components to implement in discrete form. Basically, all A/D schemes are in one way or another voltage-to-frequency conversions. The simplest form of the dual-slope DVM is shown in block diagram form in Fig. 7.24. The time base accurately establishes a fixed time T_1. At the start of the conversion cycle (start of T_1), digitally controlled switches are in the following states: SW 1 closed and SW 2 open. Thus the dc input is applied to the integrator, which had been reset to zero just prior to the start of the conversion. If the input is positive, the integrator output ramps negative, as shown in Fig. 7.25. The logic circuit maintains the state of the switches until T_1 is over, at which time SW 1 is opened and SW 2 is closed. At time T_1, the integrator output is at $-V_M$:

$$V_M = \frac{V_i T_1}{RC}$$

With SW 2 closed, the integrator input is $-V_R$, a fixed voltage. The output now ramps back toward zero. The time T_2 required for the integrator to reach zero is established by the logic block starting when T_1 is over and ending when the zero-crossing detector signals that the integrator output is zero. During T_2, the gating signal is high, thus enabling the clock pulses to enter the digital counter. The count at the end of T_2 is directly proportional to T_2. Also, since V_R and the RC product are constant, T_2 is directly proportional to V_M, which in turn is proportional to V_i:

$$T_2 = V_i \frac{T_1}{V_R}$$

At the end of T_2, the readout latches are enabled, the count contained in the counter is transferred and stored, and the counter is reset to zero. The readout displays the count which is directly proportional to the input voltage. The clock frequency, the values of R, C, and V_R, are conversion constants, which are adjusted to give a readout which is a numerical indication of the input voltage. For example, for a 1 V input, the readout would be 1000, indicating that 1000 clock pulses entered the counter during T_2. The accuracy of T_1 and V_R determines the accuracy of the conversion. Note that only short-term stability is required of R and C, i.e. their values need to remain constant only during the ramping down and then back of the integrator. Typically, T_1 is established by digital means—a crystal-controlled clock is counted down to provide the predetermined interval T_1.

Thus the basic idea of the dual-slope converter is to integrate the unknown

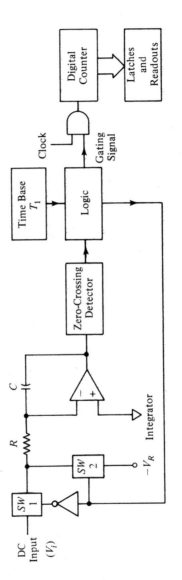

FIGURE 7.24   BASIC DUAL-SLOPE DVM BLOCK DIAGRAM

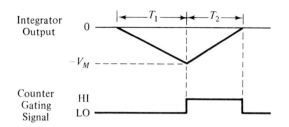

FIGURE 7.25   WAVESHAPES FOR DUAL-SLOPE AID CONVERSION

input for a known time (T_1), then integrate a known voltage (V_R) to provide a time (T_2) which is a direct measure of the input voltage. As shown in Fig. 7.25, during the first integration, the slope is variable but the integration time is fixed; during the second integration, the slope is constant but the time is variable.

The basic problems in the dual-slope conversion as outlined above are: first, the accuracy really hinges on ideal integrators and comparators; secondly, the conversion is unipolar, i.e. the scheme outlined above only converts positive voltages and cannot handle negative ones. The latter drawback can be eliminated by sensing the polarity of the input and switching in a positive or negative reference voltage V_R, which causes the integrator output to ramp back to zero. This feature is termed auto-polarity. The first problem exhibits itself in the voltage offsets of the linear blocks, the integrator, the comparator, and any buffer amplifiers that might be used. It is overcome by incorporating a separate auto-zero interval, distinct from the measure cycle. Numerous schemes exist which compensate for the offset of the analog system; all of them involve measurement of the offset and subsequent subtraction of the offset during the actual conversion cycle.

As one example of a monolithic DVM chip, consider the Siliconix LD130 shown in Fig. 7.26. This IC is an auto-zero, auto-polarity 3-digit DVM chip (in an 18-pin dip) which requires a minimum number of external components. It will convert a dc voltage in the range of $-.999$ to $+.999$ V to digital form (BCD). To decrease the number of pins, the digital output is multiplexed. In addition, underrange (less than 80 mV) and overrange conditions are decoded to allow the implementation of a fully autoranging DVM with the addition of a few ICs.

The conversion scheme used is termed *digitized feedback*, but is really only a modification of the dual-slope technique. There are basically two cycles: the auto-zero cycle and the measure cycle. During the auto-zero cycle, the AZ switch is closed, the M/Z switch is connected to ground, and the control logic causes the U/D switch to alternate between the 2 V reference signal applied to pin 2 and ground with 50% duty cycle. The effect is to charge C_{AZ} to -1 V plus or minus the net offset voltage of the whole loop (the input buffer, reference buffer, integrator, and auto-zero buffer). Thus the error due to amplifier offsets is stored on the auto-zero capacitor, and will be used during the measure cycle to cancel these

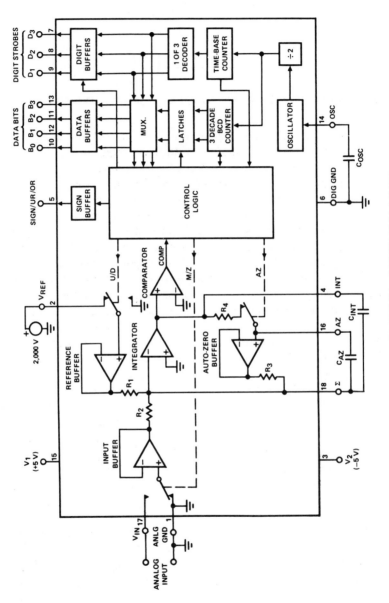

FIGURE 7.26 SILICONIX ±3 DIGIT DVM (LD130)

offsets. The -1 V results because $R_1 = R_3$ and a 50% duty cycle is used ($+2$ V and 0 V alternately).

Thus at the initiation of the measure cycle, the voltage stored on C_{AZ} and therefore applied to the integrator through R_3 is -1 V together with any offsets. When the measure cycle is initiated, the AZ switch is opened, and the input is applied to the input buffer through the M/Z switch. During the measure cycle, there are three inputs to the integrator: the input through R_2, $+2$ V or ground through R_1, and the auto-zero voltage through R_3. Determined by the state of the comparator, the control logic establishes one or two possible duty cycles for the up/down control which determines whether the 2 V reference or ground is applied to the reference buffer. Each cycle consists of eight clock pulses—one cycle gives a high up/down control for one clock pulse and a low for seven; the other cycle provides a high control for seven clock pulses and a low for one clock pulse. The state of the comparator is sampled during the seventh clock pulse; the U/D control then applies the appropriate cycle in order to return the integrator output to zero. During this time, the BCD counter is incremented or decremented by the appropriate number of clock pulses as determined by the U/D control: the count either increments by six (one down, seven up) or decrements by six (one up, seven down).

Consider a positive input (it must be below .999 V to be converted). With the reference buffer connected to ground, the integrator output must ramp positive since the input together with the auto-zero voltage provides an effective input that is negative. The integrator is forced back toward zero during the next cycle by applying the 2 V reference to the reference buffer. In this manner the net number of clock pulses in the BCD counter is a direct measure of the input voltage. (The scheme outlined above always gives a net count which is a multiple of 6; therefore, there is a short override period at the end of the conversion to allow the integrator to reach exactly zero and thus provide the desired resolution in the count.) The sign of the input ($+$ or $-$) is decoded from the presence of two consecutive timing cycles—one up, seven down or one down, seven up.

The complete ± 3 digit auto-zero DVM circuit using the LD130 is shown in Fig. 7.27. The 2 V reference is achieved by the CR033 current regulator and trimmed by the 2 k potentiometer. The BCD output is decoded by the 74C48 7-segment decoder and applied to the three LED readouts. The multiplexing signal is applied to the common-cathodes of the LED readouts through three MPS A13 *NPN* transistors—the negative sign is indicated by the MV5025 discrete LED. In order to protect the converter chip from input overvoltage, it is advisable to insert a 1 M resistor in series with the input. In addition, the analog ground, which is the input return line, must be separated from the digital ground. This is essential to the proper functioning of the converter since any switching noise on the analog ground line would provide erratic readings. This isolation can be accomplished by making certain that a separate pc trace or wire is connected from the power supply ground to pin 1, with no other connections made to this particular line other than the return

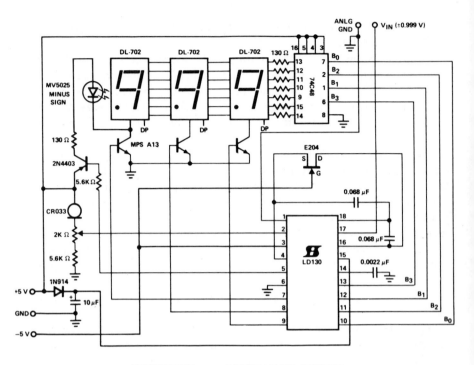

FIGURE 7.27   ±3 DIGIT LD130 CIRCUIT

for the input signal. In this manner, no digital (switching) currents flow through the analog ground line, and switching transients are essentially eliminated. If orange LED readouts are used, MAN3640, for example, the whole system depicted in Fig. 7.27 draws only a few milliamps, so that battery operation can easily be implemented. For additional information regarding the selection of capacitors as well as implementation of a complete auto-ranging DVM, see the manufacturer's data sheet in the Appendix.

Consider next the auto-zero auto-polarity 4½ digit DVM implementation shown in Fig. 7.28. This is another example of a modified dual-slope technique and uses the National Semiconductor MM5330 digital processor and LF11300 analog processor chips. The *NAND* gates labelled with a single letter are 74C00, while the two-letter *NAND* gates are 7400; the inverters are Schmitt trigger 74C14. The LH0070 is a precision 10.000 V buffered reference, derived from a temperature-compensated zener (National Semiconductor). The analog block, LF11300, is shown in Fig. 7.29; it contains buffer amplifiers, an integrator, and a comparator, as well as analog switches (S_1 through S_9). The integrating resistor and all capacitors are external.

There are two modes of operation: the auto-zero mode and the measure mode.

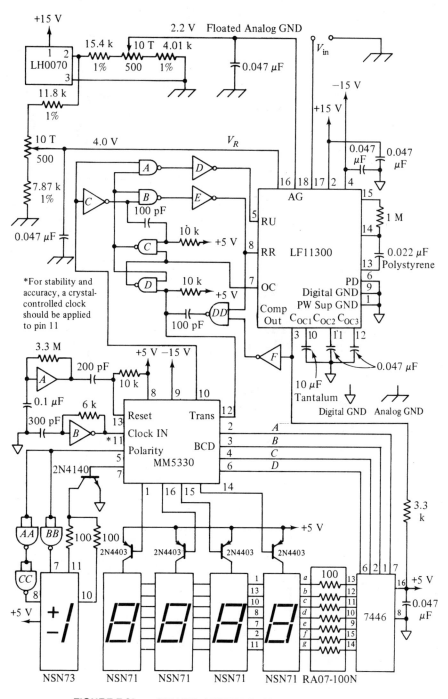

FIGURE 7.28 LF11300, MM5330 DVM APPLICATION

225

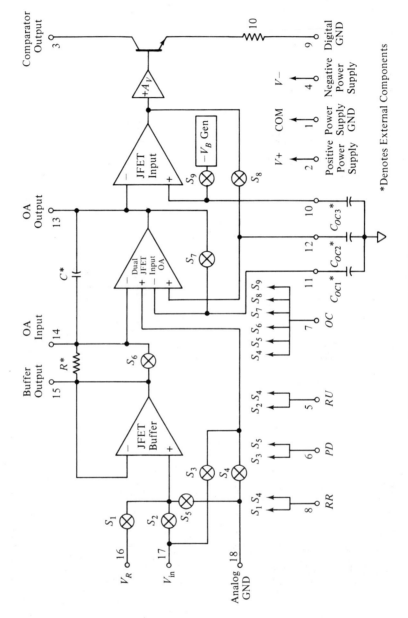

FIGURE 7.29 LFH300 SCHEMATIC

226

During the auto-zero mode, switches S_4 through S_9 are closed, so the system is effectively grounded. However, the analog ground is actually at 2.2 V. The reason for this will be apparent shortly. With the system grounded, the offsets of the buffer, integrator, and comparator are stored on C_{OC1} and C_{OC2}. This value is then used to cancel the offset-generated error during the measure cycle. In addition, C_{OC3} is used to store $-V_B$ (typically -12 V), which is the comparator reference voltage. At the conclusion of the auto-zero cycle, the measure cycle is initiated: switches S_4 through S_9 are opened; S_2 and S_4 are closed, applying the input to the buffer and subsequently to the integrator. The input voltage range is between -1.999 V and $+1.999$ V. However, since the analog ground is floated at $+2.2$ V, any input voltage will appear to the integrator as negative, thus causing it to ramp in the positive direction.

The fixed integration time of the input voltage, T_1, is established by the digital block, the MM5330 shown in Fig. 7.30. The counter internal to the MM5330 is used to provide the fixed integration time, T_1. When this time is completed, the MM5330 provides the RR (ramp reference) signal to the analog section, causing S_1 and S_4 to be closed in the LF11300 and thus the known reference voltage of 4.000 V is applied to the integrator. The output of the integrator now ramps down toward the comparator reference, $-V_B$ as shown in Fig. 7.31. The time required to reach $-V_B$, T_2, is again determined and stored in the counters of the MM5330. The count is not exactly representative of T_2, at least not in the normal sense; this is due to the 2.2 V offset of the analog ground. Note that an input of exactly zero volts produces a count of 40,000. The first 18,000 counts are used for the fixed time T_1. If zero is applied, the integrator is actually integrating the 2.2 V offset, thus giving a count of 40,000 (18,000 + 22,000). Thus, prior to decoding the BCD counter outputs and applying them to the readout, a code conversion must be carried out: a count of 40,000 is converted to 0000 at the readout. A positive input between zero and $+1.999$ V results in a counter output of between 40,000 and 60,000; this is converted to the appropriate reading (0 to $+1.999$) on the display.

A negative input between zero and -1.999 V results in a counter output of between 20,000 and 40,000, and is converted to the readout as -1.999 V to 0 V. Thus we see the use of the 2.200 V offset—it allows bipolar ($+$ and $-$) voltage conversion without the need for positive and negative reference switching. Since the input is offset by an amount larger than the highest anticipated negative input voltage, all inputs appear positive as far as the integrator is concerned. Therefore a single (4 V) reference is required to drive the integrator output back to the comparator threshold. This eliminates any imbalance between two reference voltages, one positive and one negative, that would otherwise have to be used.

Overrange signals for both positive and negative inputs are provided by displaying the polarity and a 1 in the readout of the most significant digit, with the other digits blanked. This is accomplished in the following manner. Should the integrator output cross the comparator threshold prior to a count of 20,000, this indicates that the input is more negative than -2 V (negative overrange). Should

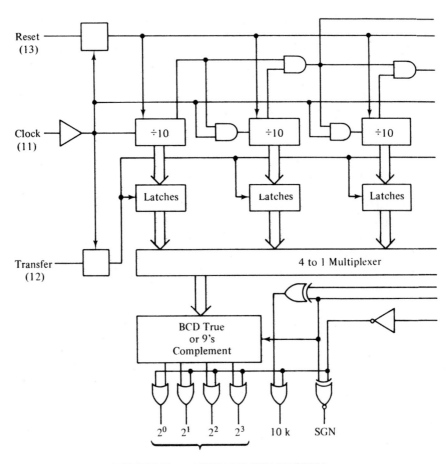

V_{SS} V_{DD}

FIGURE 7.30   MM5330 BLOCK DIAGRAM

the integrator output cross the comparator threshold after 60,000 counts, a positive input in excess of $+2$ V is indicated (positive overrange).

The circuit in Fig. 7.28 can then be used to implement a 2½ or 3½ digit DVM that is auto-zeroing as well as having the auto-polarity feature. Both the analog and digital ICs have the capability of 4½ digits; however, in order to accurately measure to the nearest tenth of a millivolt, two changes in the circuit shown in Fig.

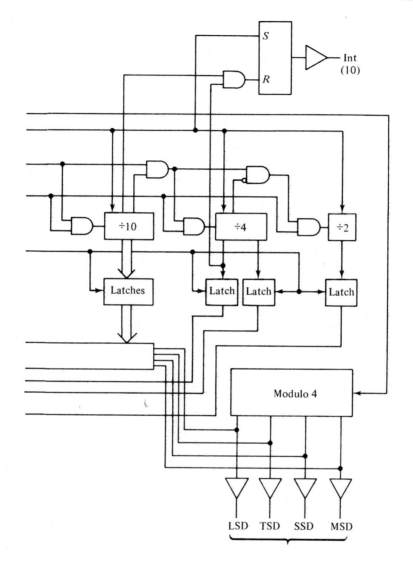

7.29 are required. First, the reference voltages needed must be stable with temperature, so that once the calibration adjustments of 2.2000 V and 4.0000 V are made, they will not drift. This is accomplished by substituting an LM399 for the LH0070. The LM399 has a 7 V zener on a thermally regulated substrate and provides the needed thermal stability. (Note that the resistors must be adjusted accordingly since the LH0070 is a 10 V reference, while the LM399 is a 7 V

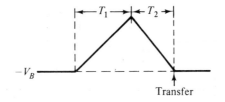

FIGURE 7.31   INTEGRATOR OUTPUT WAVESHAPE

reference.) The second change involves replacing the clock (oscillator), formed by the B Schmitt trigger and the 6 k resistor, and the 300 pF capacitor by a crystal-controlled oscillator of typically 250 kHz.

In implementing a DVM using the LF11300 and MM5330 ICs, component layout is critical. The analog and digital grounds must be separate (as discussed previously for the LD130). In addition, well-regulated supplies bypassed at the individual chips are advised. The integrating capacitor and resistor should have at least good short-term drift characteristics (the capacitor should be polystyrene or mylar, and the resistor should be metal-film or wirewound).

In the two examples discussed here, the first involving the Siliconix LD130 and the second the National MM5330-LF11300 combination, we have by no means exhausted all such chips. In fact, the two should not even be compared with one another since they obviously have different capabilities in terms of resolution and circuit complexity required. They are, however, good examples of the kind of capability which is readily available to the user. They can be used with appropriate converters to measure resistance, current, ac voltage, temperature, pressure, etc. Although Siliconix has other DVM chips (the LD110-111 for example), National LF11300 can be used in conjunction with MM5863 and 74C926, while the MM5300 can be used with discrete OP AMPs and comparators; the two specific examples chosen here serve to illustrate the following points. The LD130 provides basic DVM capability with only a handful of additional components, while the MM5300-LF11300 combination offers amazing accuracy and resolution in integrated form. The two techniques for converting an input voltage to a digital readout are also of interest themselves since they do employ a great deal of ingenuity and reflect two quite different approaches to the same problem. Finally, a number of other manufacturers (Intersil and Motorola, for example) have products which are similar.

To summarize, this section has dealt with a few examples of the power offered by relatively inexpensive and readily available ICs. In reality, the specific examples discussed represent only the tip of the iceberg—they only begin to indicate those devices that are already available now and those devices that are soon to be available.

REVIEW QUESTIONS

7.1   When a 555 timer is used as a one-shot, what is the steady-state or untriggered input and output if the supply is 5 V?

7.2   In order to trigger a 555 timer in the one-shot configuration, what must be the condition on pin 2 and for how long, if the supply voltage is 10 V?

7.3   Specify the timing components (resistor and capacitor) to provide a 100 μs pulse width using a 555 timer.

7.4   What is the maximum range of pulse width (delay) that can be obtained using a single 555 timer as a one-shot? How can longer pulse widths be obtained?

7.5   Use a 555 timer as shown in Fig. 7.5 to provide a system clock (source of square waves) with a frequency of 50 kHz and a duty cycle of 30%. Specify the circuit components.

7.6   Use a 555 in the one-shot configuration to convert the output of the timer in Question 7.5 above into a 50% duty-cycle square wave at 50 kHz. Specify the timing components.

7.7   In applications requiring a voltage-controlled oscillator, why would it be more desirable to use a 566 rather than a 555?

7.8   When a 566 is used as a VCO with a supply voltage of 10 V, to obtain a frequency change factor of between 1 and 10, what must be the range of the input voltage? (Specify numerical limits.)

7.9   The 566 is used to provide a frequency-modulated output as shown in Fig. 7.12. The frequency with no input applied is 10 kHz. The input signal is a 1 V peak-to-peak triangle wave. If the supply voltage is 12 V, determine the limits in the output frequency variation.

7.10 In Question 7.9, what are the expected amplitudes for the triangle-wave and square-wave outputs?

7.11 How could a 566 VCO be used to provide a triangle wave which has an average or dc level of zero volts?

7.12 If an 8038 function generator is used with a 9 V supply, specify the peak voltage levels for all three possible outputs. What are the minimum voltage levels for the three outputs?

7.13 Consider the linearly voltage-controlled function generator shown in Fig. 7.17. How do potentiometers P_1 and P_3 set the symmetry of the sine and triangle outputs? Why are separate high- and low-frequency symmetry adjustments necessary? (Consult Fig. 7.13 for the circuit diagram of the 8038.)

7.14   If the circuit shown in Fig. 7.17 were implemented with ± 12 V supplies, specify the range of V_{in}, the amplitude of the sine- and square-wave outputs and the voltage limits on the triangle-wave output.

7.15   Show a block diagram application of an XR2206 to provide a sine-wave output which is amplitude modulated.

7.16   In Question 7.15 above, a 12 V supply is used, and the modulation should be between 30 and 90%, with an output frequency selectable between 30 and 50 kHz. Specify all external circuit values as well as the amplitude requirements for the modulating signal.

7.17   It is desired to implement a swept-frequency oscillator using an XR2206. The oscillator is to have two ranges (selectable by either a switch or a logic TTL input): 1 kHz to 10 kHz and 10 kHz to 100 kHz. In all cases, the sweep rate is to be 100 ms, i.e. the range of output frequencies is to be repeated 10 times each second. Specify all external components and connections. (*Hint*: Use the circuit in Fig. 7.23 to generate the sweep voltage.)

7.18   Describe the basic dual-slope conversion as used in most digital voltmeters.

7.19   Describe the modified dual-slope conversion as used in most digital voltmeters.

7.20   Describe the modified dual-slope conversion employed in the voltmeter application utilizing the LD130 chip.

7.21   How are auto-zero and auto-polarity functions accomplished in the LD130?

7.22   Describe the modified dual-slope conversion as used in the LF11300-MM5330 application for a digital voltmeter.

7.23   How are auto-zero and auto-polarity functions accomplished in the LF11300-MM5330 application?

Appendix

Manufacturers' Data Sheets

This appendix contains a small selection of manufacturers' data sheets—those for some of the specific devices discussed in the previous seven chapters. The intent here is only to convey the appearance and typical content of data sheets. The manufacturers all provide data sheets or data books fully describing their products. In addition, they make available application notes and booklets. Therefore, the user is advised to obtain the most up-to-date information regarding any IC directly from the manufacturer. To that end, a partial list of IC manufacturers with their mailing addresses is given here:

Fairchild Semiconductor
464 Ellis Street
Mountain View, California 94042

Intersil, Incorporated
10900 North Tantau Avenue
Cupertino, California 95014

Motorola, Incorporated
Semiconductor Products Div.
5005 East McDowell Road
Phoenix, Arizona 85008

National Semiconductor Corp.
2900 Semiconductor Drive
Santa Clara, California 95051

RCA Solid State
Box 3200
Sommerville, New Jersey 08876

Signetics Corporation
811 East Arques Avenue
Sunnyvale, California 94086

Siliconix, Incorporated
2201 Laurelwood Road
Santa Clara, California 95054

Texas Instruments, Inc.
P.O. Box 5012
Dallas, Texas 75222

LM741/LM741C operational amplifier[*]

general description

The LM741 and LM741C are general purpose operational amplifiers which feature improved performance over industry standards like the LM709. They are direct, plug-in replacements for the 709C, LM201, MC1439 and 748 in most applications.

The offset voltage and offset current are guaranteed over the entire common mode range. The amplifiers also offer many features which make their application nearly foolproof: overload protection on the input and output, no latch-up when the common mode range is exceeded, as well as freedom from oscillations.

The LM741C is identical to the LM741 except that the LM741C has its performance guaranteed over a 0°C to 70°C temperature range, instead of –55°C to 125°C.

schematic and connection diagrams

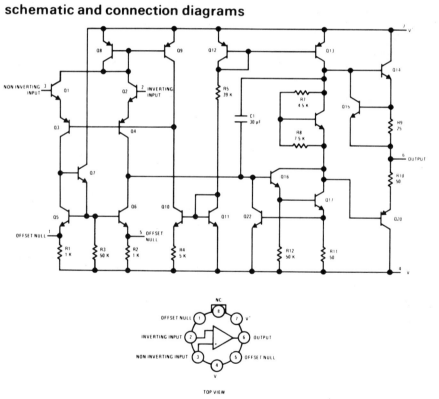

TOP VIEW

NOTE Pin 4 connected to case

Courtesy of National Semiconductor Corporation

235

LM741/741C

absolute maximum ratings

Supply Voltage LM741		±22V
LM741C		±18V
Power Dissipation (Note 1)		500 mW
Differential Input Voltage		±30V
Input Voltage (Note 2)		±15V
Output Short-Circuit Duration		Indefinite
Operating Temperature Range LM741		-55°C to 125°C
LM741C		0°C to 70°C
Storage Temperature Range		-65°C to 150°C
Lead Temperature (Soldering, 10 sec)		300°C

electrical characteristics (Note 3)

PARAMETER	CONDITIONS	LM741			LM741C			UNITS
		MIN	TYP	MAX	MIN	TYP	MAX	
Input Offset Voltage	$T_A = 25^\circ$C, $R_S < 10$ kΩ		1.0	5.0		1.0	6.0	mV
Input Offset Current	$T_A = 25^\circ$C		30	200		30	200	nA
Input Bias Current	$T_A = 25^\circ$C		200	500		200	500	nA
Input Resistance	$T_A = 25^\circ$C	0.3	1.0		0.3	1.0		MΩ
Supply Current	$T_A = 25^\circ$C, $V_S = \pm15$V		1.7	2.8		1.7	2.8	mA
Large Signal Voltage Gain	$T_A = 25^\circ$C, $V_S = \pm15$V $V_{OUT} = \pm10$V, $R_L > 2$ kΩ	50	160		25	160		V/mV
Input Offset Voltage	$R_S < 10$ kΩ			6.0			7.5	mV
Input Offset Current				500			300	nA
Input Bias Current				1.5			0.8	µA
Large Signal Voltage Gain	$V_S = \pm15$V, $V_{OUT} = \pm10$V $R_L > 2$ kΩ	25			15			V/mV
Output Voltage Swing	$V_S = \pm15$V, $R_L = 10$ kΩ	±12	±14		±12	±14		V
	$R_L = 2$ kΩ	±10	±13		±10	±13		V
Input Voltage Range	$V_S = \pm15$V	±12			±12			V
Common Mode Rejection Ratio	$R_S < 10$ kΩ	70	90		70	90		dB
Supply Voltage Rejection Ratio	$R_S < 10$ kΩ	77	96		77	96		dB

Note 1: The maximum junction temperature of the LM741 is 150°C, while that of the LM741C is 100°C. For operating at elevated temperatures, devices in the TO-5 package must be derated based on a thermal resistance of 150°C/W, junction to case.

Note 2: For supply voltages less than ±15V, the absolute maximum input voltage is equal to the supply voltage.

Note 3: These specifications apply for $V_S = \pm15$V and -55°C $\leq T_A \leq 125^\circ$C, unless otherwise specified. With the LM741C, however, all specifications are limited to 0°C $\leq T_A \leq 70^\circ$C and $V_S = \pm15$V.

LM4250/LM4250C
programmable operational amplifier[*]

general description

The LM4250 and LM4250C are extremely versatile programmable monolithic operational amplifiers. A single external master bias current setting resistor programs the input bias current, input offset current, quiescent power consumption, slew rate, input noise, and the gain-bandwidth product. The device is a truly general purpose operational amplifier.

features

- ±1V to ±18V power supply operation
- 3 nA input offset current

- Standby power consumption as low as 500 nW
- No frequency compensation required
- Programmable electrical characteristics
- Offset Voltage nulling capability
- Can be powered by two flashlight batteries
- Short circuit protection

The LM4250C is identical to the LM4250 except that the LM4250C has its performance guaranteed over a 0°C to 70°C temperature range instead of the −55°C to +125°C temperature range of the LM4250.

schematic and connection diagrams

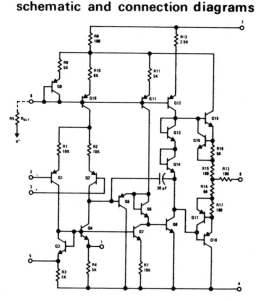

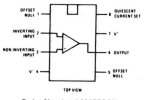

Metal Can Package

Order Number LM4250H or LM4250CH
See Package 11

Dual-In-Line Package

Order Number LM4250CN
See Package 20

typical applications

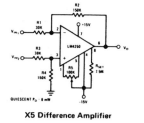

X5 Difference Amplifier

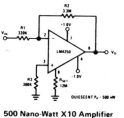

500 Nano-Watt X10 Amplifier

Courtesy of National Semiconductor Corporation

237

LM4250/LM4250C

absolute maximum ratings

Supply Voltage	±18V	Output Short-Circuit Duration	Indefinite
Power Dissipation (Note 1)	500 mW	Operating Temperature Range LM4250	$-55°C \leq T_A \leq 125°C$
Differential Input Voltage	±30V	LM4250C	$0°C \leq T_A \leq 70°C$
Input Voltage (Note 2)	±15V	Storage Temperature Range	$-65°C$ to $150°C$
I_{SET} Current	150 µA	Lead Temperature (Soldering, 10 sec)	$300°C$

electrical characteristics LM4250 ($-55°C \leq T_A \leq 125°C$ unless otherwise specified)

PARAMETERS	CONDITIONS	$V_S = \pm1.5V$			
		$I_{SET} = 1\,\mu A$		$I_{SET} = 10\,\mu A$	
		MIN	MAX	MIN	MAX
V_{OS}	$T_A = 25°$ $R_S \leq 100\,k\Omega$		3 mV		5 mV
I_{OS}	$T_A = 25°$		3 nA		10 nA
I_{bias}	$T_A = 25°$		7.5 nA		50 nA
Large Signal Voltage Gain	$T_A = 25°$ $R_L = 100\,k\Omega$	40k			
	$V_O = \pm0.6$, $R_L = 10\,k\Omega$			50k	
Supply Current	$T_A = 25°C$		7.5 µA		80 µA
Power Consumption	$T_A = 25°C$		23 µW		240 µW
V_{OS}	$R_S \leq 100\,k\Omega$		4 mV		6 mV
I_{OS}	$T_A = 125°C$		5 nA		10 nA
	$T_A = -55°C$		3 nA		10 nA
I_{bias}			7.5 nA		50 nA
Input Voltage Range		±0.7V		±0.7V	
Large Signal Voltage Gain	$V_O = \pm0.6V$ $R_L = 100\,k\Omega$	30k			
	$R_L = 10\,k\Omega$			30k	
Output Voltage Swing	$R_L = 100\,k\Omega$		±0.6V		
	$R_L = 10\,k\Omega$			±0.6V	
Common Mode Rejection Ratio	$R_S \leq 10\,k\Omega$	70 dB		70 dB	
Supply Voltage Rejection Ratio	$R_S \leq 10\,k\Omega$	76 dB		76 dB	
Supply Current			8 µA		90 µA
Power Consumption			24 µW		270 µW

PARAMETERS	CONDITIONS	$V_S = \pm15V$			
		$I_{SET} = 1\,\mu A$		$I_{SET} = 10\,\mu A$	
		MIN	MAX	MIN	MAX
V_{OS}	$T_A = 25°C$ $R_S \leq 100\,k\Omega$		3 mV		5 mV
I_{OS}	$T_A = 25°C$		3 nA		10 nA
I_{bias}	$T_A = 25°C$		7.5 nA		50 nA
Large Signal Voltage Gain	$T_A = 25°C$ $R_L = 100\,k\Omega$	100k			
	$V_O = \pm10V$ $R_L = 10\,k\Omega$			100k	
Supply Current	$T_A = 25°C$		10 µA		90 µA
Power Consumption	$T_A = 25°C$		300 µW		2.7 mW
V_{OS}	$R_S \leq 100\,k\Omega$		4 mV		6 mV
I_{OS}	$T_A = 125°C$		25 nA		25 nA
	$T_A = -55°C$		3 nA		10 nA
I_{bias}			7.5 nA		50 nA
Input Voltage Range		±13.5V		±13.5V	
Large Signal Voltage Gain	$V_O = \pm10V$ $R_L = 100\,k\Omega$	50k			
	$R_L = 10\,k\Omega$			50k .	
Output Voltage Swing	$R_L = 100\,k\Omega$	±12V			
	$R_L = 10\,k\Omega$			±12V	
Common Mode Rejection Ratio	$R_S \leq 10\,k\Omega$	70 dB		70 dB	
Supply Voltage Rejection Ratio	$R_S \leq 10\,k\Omega$	76 dB		76 dB	
Supply Current			11 µA		100 µA
Power Consumption			330 µW		3 mW

Note 1: The maximum junction temperature of the LM4250 is $150°C$, while that of the LM4250C is $100°C$. For operating at elevated temperatures, devices in the TO-5 package must be derated based on a thermal resistance of $150°C/W$ junction to ambient, or $45°C/W$ junction to case. The thermal resistance of the dual-in-line package is $125°C/W$.

Note 2: For supply voltages less than ±15V, the absolute maximum input voltage is equal to the supply voltage.

LM4250/LM4250C

electrical characteristics LM4250C (0°C ≤ T_A ≤ 70°C unless otherwise specified)

PARAMETERS	CONDITIONS	V_S = ±1.5V			
		I_{SET} = 1 μA		I_{SET} = 10 μA	
		MIN	MAX	MIN	MAX
V_{OS}	T_A = 25°C R_S ≤ 100 kΩ		5 mV		6 mV
I_{OS}	T_A = 25°C		6 nA		20 nA
I_{bias}	T_A = 25°C		10 nA		75 nA
Large Signal Voltage Gain	T_A = 25°C R_L = 100 kΩ	25k			
	V_O = ±0.6V R_L = 10 kΩ			25k	
Supply Current	T_A = 25°C		8 μA		90 μA
Power Consumption	T_A = 25°C		24 μW		270 μW
V_{OS}	R_S ≤ 10 kΩ		6.5 mV		7.5 mV
I_{OS}			8 nA		25 nA
I_{bias}			10 nA		80 nA
Input Voltage Range		±0.6V		±0.6V	
Large Signal Voltage Gain	V_O = ±0.6V R_L = 100 kΩ	25k			
	R_L = 10 kΩ			25k	
Output Voltage Swing	R_L = 100 kΩ	±0.6V			
	R_L = 10 kΩ			±0.6V	
Common Mode Rejection Ratio	R_S ≤ 10 kΩ	70 dB		70 dB	
Supply Voltage Rejection Ratio	R_S ≤ 10 kΩ	74 dB		74 dB	
Supply Current			8 μA		90 uA
Power Consumption			24 μW		270 uW

PARAMETERS	CONDITIONS	V_S = ±15V			
		I_{SET} = 1 μA		I_{SET} = 10 μA	
		MIN	MAX	MIN	MAX
V_{OS}	T_A = 25°C R_S ≤ 100 kΩ		5 mV		6 mV
I_{OS}	T_A = 25°C		6 nA		20 nA
I_{bias}	T_A = 25°C		10 nA		75 nA
Large Signal Voltage Gain	T_A = 25°C R_L = 100 kΩ	60k			
	V_O = ±10V R_L = 10 kΩ			60k	
Supply Current	T_A = 25°C		11 μA		100 μA
Power Consumption	T_A = 25°C		330 μW		3 mW
V_{OS}	R_S ≤ 10 kΩ		6.5 mV		7.5 mV
I_{OS}			8 nA		25 nA
I_{bias}			10 nA		80 nA
Input Voltage Range		±13.5V		±13.5V	
Large Signal Voltage Gain	V_O = ±10V R_L = 100 kΩ	50k			
	R_L = 10 kΩ			50k	
Output Voltage Swing	R_L = 100 kΩ	±12V			
	R_L = 10 kΩ			±12V	
Common Mode Rejection Ratio	R_S ≤ 10 kΩ	70 dB		70 dB	
Supply Voltage Rejection Ratio	R_S ≤ 10 kΩ	74 dB		74 dB	
Supply Current			11 uA		100 uA
Power Consumption			300 uW		3 mW

resistor biasing

Set Current Setting Resistor to V⁻

V_S	I_{SET}				
	0.1 μA	0.5 μA	1.0 μA	5 μA	10 μA
±1.5V	25.6 MΩ	5.04 MΩ	2.5 MΩ	492 kΩ	244 kΩ
±3.0V	55.6 MΩ	11.0 MΩ	5.5 MΩ	1.09 MΩ	544 kΩ
±6.0V	116 MΩ	23.0 MΩ	11.5 MΩ	2.29 MΩ	1.14 MΩ
±9.0V	176 MΩ	35.0 MΩ	17.5 MΩ	3.49 MΩ	1.74 MΩ
±12.0V	236 MΩ	47.0 MΩ	23.5 MΩ	4.69 MΩ	2.34 MΩ
±15.0V	296 MΩ	59.0 MΩ	29.5 MΩ	5.89 MΩ	2.94 MΩ

LM4250/LM4250C

typical performance characteristics

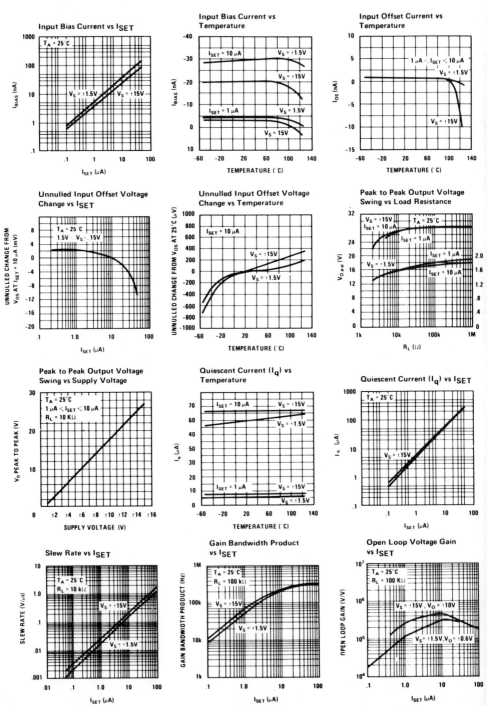

LM4250/LM4250C

typical performance characteristics (con't)

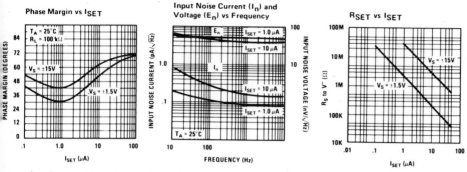

Phase Margin vs I_{SET}

Input Noise Current (I_n) and Voltage (E_n) vs Frequency

R_{SET} vs I_{SET}

typical applications (con't)

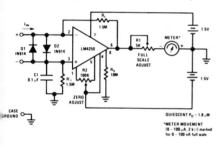

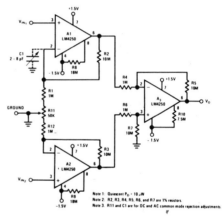

X100 Instrumentation Amplifier 10 μW

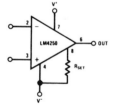

Floating Input Meter Amplifier
100 Nano-Ampere Full Scale

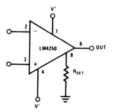

R_{SET} Connected to V⁻

R_{SET} Connected to Ground

I_{SET} EQUATIONS:

$$I_{SET} \approx \frac{V^+ + |V^-| - 0.5}{R_{SET}} \quad \text{where } R_{SET} \text{ is connected to } V^-.$$

$$I_{SET} \approx \frac{V^+ - 0.5}{R_{SET}} \quad \text{where } R_{SET} \text{ is connected to ground.}$$

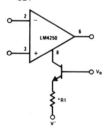

Transistor Current Source Biasing

*R1 limits I_{SET} maximum

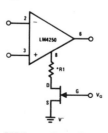

FET Current Source Biasing

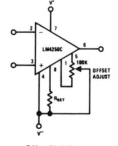

Offset Null Circuit

241

LM723/LM723C voltage regulator[*]

general description

The LM723/LM723C is a voltage regulator designed primarily for series regulator applications. By itself, it will supply output currents up to 150 mA; but external transistors can be added to provide any desired load current. The circuit features extremely low standby current drain, and provision is made for either linear or foldback current limiting. Important characteristics are:

- 150 mA output current without external pass transistor
- Output currents in excess of 10A possible by adding external transistors

- Input voltage 40V max
- Output voltage adjustable from 2V to 37V
- Can be used as either a linear or a switching regulator.

The LM723/LM723C is also useful in a wide range of other applications such as a shunt regulator, a current regulator or a temperature controller.

The LM723C is identical to the LM723 except that the LM723C has its performance guaranteed over a 0°C to 70°C temperature range, instead of -55°C to +125°C.

schematic and connection diagrams [*]

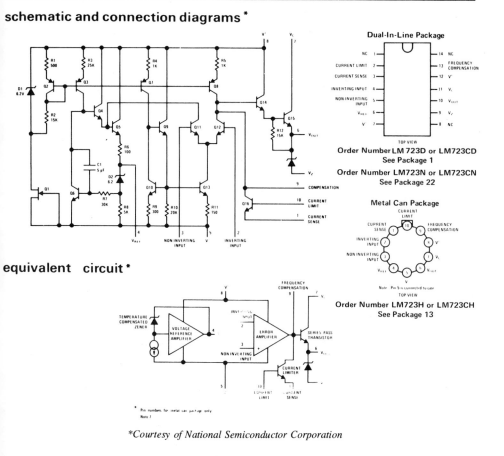

Order Number LM 723D or LM723CD
See Package 1

Order Number LM723N or LM723CN
See Package 22

Order Number LM723H or LM723CH
See Package 13

Courtesy of National Semiconductor Corporation

243

LM723/LM723C

absolute maximum ratings

Pulse Voltage from V^+ to V^- (50 ms)	50V
Continuous Voltage from V^+ to V^-	40V
Input-Output Voltage Differential	40V
Maximum Amplifier Input Voltage (Either Input)	7.5V
Maximum Amplifier Input Voltage (Differential)	5V
Current from V_Z	25 mA
Current from V_{REF}	15 mA
Internal Power Dissipation Metal Can (Note 1)	800 mW
Cavity DIP (Note 1)	900 mW
Molded DIP (Note 1)	660 mW
Operating Temperature Range LM723	$-55°C$ to $+125°C$
LM723C	$0°C$ to $+70°C$
Storage Temperature Range Metal Can	$-65°C$ to $+150°C$
DIP	$-55°C$ to $+125°C$
Lead Temperature (Soldering, 10 sec)	$300°C$

electrical characteristics (Note 2)

PARAMETER	CONDITIONS	LM723			LM723C			UNITS
		MIN	TYP	MAX	MIN	TYP	MAX	
Line Regulation	V_{IN} = 12V to V_{IN} = 15V		.01	0.1		.01	0.1	% V_{OUT}
	$-55°C < T_A < +125°C$			0.3				% V_{OUT}
	$0°C < T_A < +70°C$						0.3	% V_{OUT}
	V_{IN} = 12V to V_{IN} = 40V		.02	0.2		0.1	0.5	% V_{OUT}
Load Regulation	I_L = 1 mA to I_L = 50 mA		.03	0.15		.03	0.2	% V_{OUT}
	$-55°C < T_A < +125°C$			0.6				%V_{OUT}
	$0°C < T_A < = +70°C$						0.6	%V_{OUT}
Ripple Rejection	f = 50 Hz to 10 kHz, C_{REF} = 0		74			74		dB
	f = 50 Hz to 10 kHz, C_{REF} = 5 μF		86			86		dB
Average Temperature	$-55°C < T_A < +125°C$		.002	.015				%/°C
Coefficient of Output Voltage	$0°C < T_A < +70°C$					.003	.015	%/°C
Short Circuit Current Limit	R_{SC} = 10Ω, V_{OUT} = 0		65			65		mA
Reference Voltage		6.95	7.15	7.35	6.80	7.15	7.50	V
Output Noise Voltage	BW = 100 Hz to 10 kHz, C_{REF} = 0		20			20		μVrms
	BW = 100 Hz to 10 kHz, C_{REF} = 5 μF		2.5			2.5		μVrms
Long Term Stability			0.1			0.1		%/1000 hrs
Standby Current Drain	I_L = 0, V_{IN} = 30V		1.3	3.5		1.3	4.0	mA
Input Voltage Range		9.5		40	9.5		40	V
Output Voltage Range		2.0		37	2.0		37	V
Input-Output Voltage Differential		3.0		38	3.0		38	V

Note 1: See derating curves for maximum power rating above 25°C.

Note 2: Unless otherwise specified, T_A = 25°C, V_{IN} = V^+ = V_C = 12V, V^- = 0, V_{OUT} = 5V, I_L = 1 mA, R_{SC} = 0, C_1 = 100 pF, C_{REF} = 0 and divider impedance as seen by error amplifier < 10 kΩ connected as shown in Figure 1. Line and load regulation specifications are given for the condition of constant chip temperature. Temperature drifts must be taken into account separately for high dissipation conditions.

Note 3: L_1 is 40 turns of No. 20 enameled copper wire wound on Ferroxcube P36/22-3B7 pot core or equivalent with 0.009 in. air gap.

Note 4: Figures in parentheses may be used if R1/R2 divider is placed on opposite input of error amp.

Note 5: Replace R1/R2 in figures with divider shown in Figure 13.

Note 6: V^+ must be connected to a +3V or greater supply.

Note 7: For metal can applications where V_Z is required, an external 6.2 volt zener diode should be connected in series with V_{OUT}.

244

LM723/LM723C

maximum power ratings

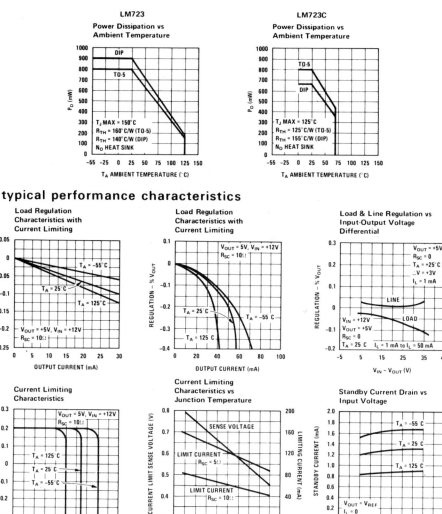

LM723
Power Dissipation vs Ambient Temperature

T_J MAX = 150°C
R_{TH} = 160°C/W (TO-5)
R_{TH} = 140°C/W (DIP)
N_O HEAT SINK

LM723C
Power Dissipation vs Ambient Temperature

T_J MAX = 125°C
R_{TH} = 125°C/W (TO-5)
R_{TH} = 155°C/W (DIP)
N_O HEAT SINK

typical performance characteristics

Load Regulation Characteristics with Current Limiting

V_{OUT} = +5V, V_{IN} = +12V
R_{SC} = 10Ω

Load Regulation Characteristics with Current Limiting

V_{OUT} = 5V, V_{IN} = +12V
R_{SC} = 10Ω

Load & Line Regulation vs Input-Output Voltage Differential

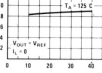

V_{OUT} = +5V
R_{SC} = 0
T_A = +25°C
ΔV = +3V
I_L = 1 mA

V_{IN} = +12V
V_{OUT} = +5V
R_{SC} = 0
T_A = 25°C I_L = 1 mA to I_L = 50 mA

Current Limiting Characteristics

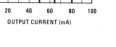

V_{OUT} = 5V, V_{IN} = +12V
R_{SC} = 10Ω

Current Limiting Characteristics vs Junction Temperature

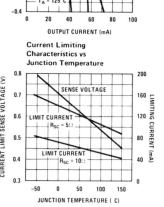

SENSE VOLTAGE
LIMIT CURRENT R_{SC} = 5Ω
LIMIT CURRENT R_{SC} = 10Ω

Standby Current Drain vs Input Voltage

T_A = -55°C
T_A = 25°C
T_A = 125°C
V_{OUT} = V_{REF}
I_L = 0

Line Transient Response

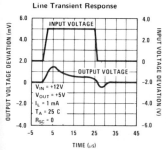

INPUT VOLTAGE
OUTPUT VOLTAGE
V_{IN} = +12V
V_{OUT} = +5V
I_L = 1 mA
T_A = 25°C
R_{SC} = 0

Load Transient Response

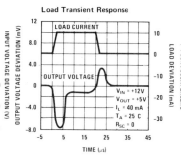

LOAD CURRENT
OUTPUT VOLTAGE
V_{IN} = +12V
V_{OUT} = +5V
I_L = 40 mA
T_A = 25°C
R_{SC} = 0

Output Impedance vs Frequency

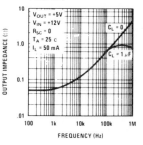

V_{OUT} = +5V
V_{IN} = +12V
R_{SC} = 0
T_A = 25°C
I_L = 50 mA
C_L = 0
C_L = 1 μF

TABLE I   RESISTOR VALUES (kΩ) FOR STANDARD OUTPUT VOLTAGE

POSITIVE OUTPUT VOLTAGE	APPLICABLE FIGURES	FIXED OUTPUT ±5%		OUTPUT ADJUSTABLE ±10% (Note 5)			NEGATIVE OUTPUT VOLTAGE	APPLICABLE FIGURES	FIXED OUTPUT ±5%		5% OUTPUT ADJUSTABLE ±10%		
		R1	R2	R1	P1	R2			R1	R2	R1	P1	R2
	(Note 4)												
+3.0	1, 5, 6, 9, 12 (4)	4.12	3.01	1.8	0.5	1.2	+100	7	3.57	102	2.2	10	91
+3.6	1, 5, 6, 9, 12 (4)	3.57	3.65	1.5	0.5	1.5	+250	7	3.57	255	2.2	10	240
+5.0	1, 5, 6, 9, 12 (4)	2.15	4.99	.75	0.5	2.2	−6 (Note 6)	3, (10)	3.57	2.43	1.2	0.5	.75
+6.0	1, 5, 6, 9, 12 (4)	1.15	6.04	0.5	0.5	2.7	−9	3, 10	3.48	5.36	1.2	0.5	2.0
+9.0	2, 4, (5, 6, 12, 9)	1.87	7.15	.75	1.0	2.7	−12	3, 10	3.57	8.45	1.2	0.5	3.3
+12	2, 4, (5, 6, 9, 12)	4.87	7.15	2.0	1.0	3.0	−15	3, 10	3.65	11.5	1.2	0.5	4.3
+15	2, 4, (5, 6, 9, 12)	7.87	7.15	3.3	1.0	3.0	−28	3, 10	3.57	24.3	1.2	0.5	10
+28	2, 4, (5, 6, 9, 12)	21.0	7.15	5.6	1.0	2.0	−45	8	3.57	41.2	2.2	10	33
+45	7	3.57	48.7	2.2	10	39	−100	8	3.57	97.6	2.2	10	91
+75	7	3.57	78.7	2.2	10	68	−250	8	3.57	249	2.2	10	240

TABLE II   FORMULAE FOR INTERMEDIATE OUTPUT VOLTAGES

Outputs from +2 to +7 volts
[Figures 1, 5, 6, 9, 12, (4)]

$$V_{OUT} = [V_{REF} \times \frac{R2}{R1 + R2}]$$

Outputs from +7 to +37 volts
[Figures 2, 4, (5, 6, 9, 12)]

$$V_{OUT} = [V_{REF} \times \frac{R1 + R2}{R2}]$$

Outputs from +4 to 250 volts
[Figure 7]

$$V_{OUT} = [\frac{V_{REF}}{2} \times \frac{R2 - R1}{R1}] ; R3 = R4$$

Outputs from −6 to −250 volts
[Figures 3, 8, 10]

$$V_{OUT} = [\frac{V_{REF}}{2} \times \frac{R1 + R2}{R1}], R3 = R4$$

Current Limiting

$$I_{LIMIT} = \frac{V_{SENSE}}{R_{SC}}$$

Foldback Current Limiting

$$I_{KNEE} = [\frac{V_{OUT} \, R3}{R_{SC} \, R4} + \frac{V_{SENSE} \, (R3 + R4)}{R_{SC} \, R4}]$$

$$I_{SHORT \, CKT} = [\frac{V_{SENSE}}{R_{SC}} \times \frac{R3 + R4}{R4}]$$

typical applications

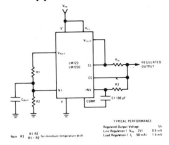

FIGURE 1. Basic Low Voltage Regulator
(V_{OUT} = 2 to 7 Volts)

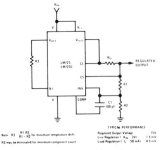

FIGURE 2. Basic High Voltage Regulator
(V_{OUT} = 7 to 37 Volts)

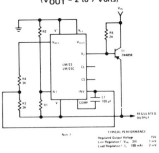

FIGURE 3. Negative Voltage Regulator

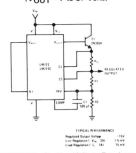

FIGURE 4. Positive Voltage Regulator
(External NPN Pass Transistor)

LM723/LM723C

typical applications (con't.)

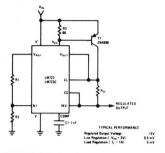

FIGURE 5. Positive Voltage Regulator
(External PNP Pass Transistor)

TYPICAL PERFORMANCE
Regulated Output Voltage +5V
Line Regulation (ΔV$_{IN}$ = 3V) 0.5 mV
Load Regulation (ΔI$_L$ = 1A) 5 mV

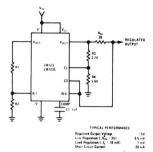

FIGURE 6. Foldback Current Limiting

TYPICAL PERFORMANCE
Regulated Output Voltage +5V
Line Regulation (ΔV$_{IN}$ = 3V) 0.5 mV
Load Regulation (ΔI$_L$ = 10 mA) 1 mV
Short Circuit Current 20 mA

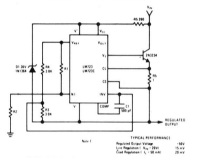

FIGURE 7. Positive Floating Regulator

TYPICAL PERFORMANCE
Regulated Output Voltage +50V
Line Regulation (V$_{IN}$ = 20V) 15 mV
Load Regulation (I$_L$ = 50 mA) 20 mV

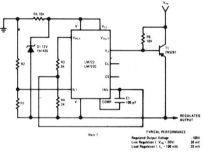

FIGURE 8. Negative Floating Regulator

TYPICAL PERFORMANCE
Regulated Output Voltage −100V
Line Regulation (V$_{IN}$ = 20V) 30 mV
Load Regulation (I$_L$ = 100 mA) 20 mV

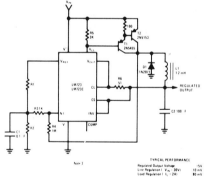

FIGURE 9. Positive Switching Regulator

TYPICAL PERFORMANCE
Regulated Output Voltage +5V
Line Regulation (V$_{IN}$ = 30V) 10 mV
Load Regulation (I$_L$ = 2A) 80 mV

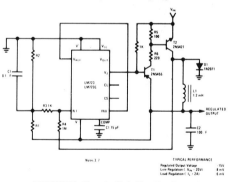

FIGURE 10. Negative Switching Regulator

TYPICAL PERFORMANCE
Regulated Output Voltage −15V
Line Regulation (V$_{IN}$ = 20V) 8 mV
Load Regulation (I$_L$ = 2A) 6 mV

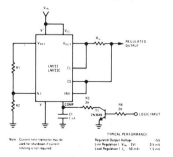

FIGURE 11. Remote Shutdown Regulator with
Current Limiting

Note Current limit transistor may be used for shutdown if current limiting is not required

TYPICAL PERFORMANCE
Regulated Output Voltage +5V
Line Regulation (V$_{IN}$ = 3V) 0.5 mV
Load Regulation (I$_L$ = 50 mA) 1.5 mV

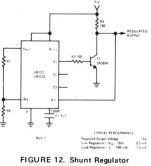

FIGURE 12. Shunt Regulator

TYPICAL PERFORMANCE
Regulated Output Voltage +5V
Line Regulation (V$_{IN}$ = 10V) 0.5 mV
Load Regulation (I$_L$ = 100 mA) 1.5 mV

247

LM309 five-volt regulator[*]

general description

The LM309 is a complete 5V regulator fabricated on a single silicon chip. It is designed for local regulation on digital logic cards, eliminating the distribution problems associated with single-point regulation. The device is available in two common transistor packages. In the solid-kovar TO-5 header, it can deliver output currents in excess of 200 mA, if adequate heat sinking is provided. With the TO-3 power package, the available output current is greater than 1A.

The regulator is essentially blow-out proof. Current limiting is included to limit the peak output current to a safe value. In addition, thermal shutdown is provided to keep the IC from overheating. If internal dissipation becomes too great, the regulator will shut down to prevent excessive heating.

Considerable effort was expended to make the LM309 easy to use and minimize the number of external components. It is not necessary to bypass the output, although this does improve transient response somewhat. Input bypassing is needed, however, if the regulator is located very far from the filter capacitor of the power supply. Stability is also achieved by methods that provide very good rejection of load or line transients as are usually seen with TTL logic.

Although designed primarily as a fixed-voltage regulator, the output of the LM309 can be set to voltages above 5V, as shown below. It is also possible to use the circuit as the control element in precision regulators, taking advantage of the good current-handling capability and the thermal overload protection.

To summarize, outstanding features of the regulator are:

- Specified to be compatible, worst case, with TTL and DTL
- Output current in excess of 1A
- Internal thermal overload protection
- No external components required

schematic diagram

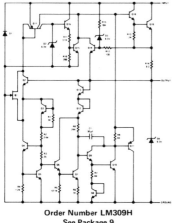

Order Number LM309H
See Package 9

Order Number LM309K
See Package 18

typical applications

High Stability Regulator*

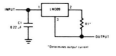

Fixed 5V Regulator

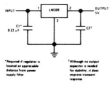

Adjustable Output Regulator

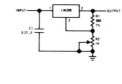

Current Regulator

Courtesy of National Semiconductor Corporation

LM309

absolute maximum ratings

Input Voltage	35V
Power Dissipation	Internally Limited
Operating Junction Temperature Range	0°C to 125°C
Storage Temperature Range	-65°C to 150°C
Lead Temperature (Soldering, 10 sec)	300°C

design characteristics (Note 1)

PARAMETER	CONDITIONS	MIN	TYP	MAX	UNITS
Output Voltage	$T_j = 25^{\circ}$C	4.8	5.05	5.2	V
Line Regulation	$T_j = 25^{\circ}$C $7V \leq V_{IN} \leq 25V$		4.0	50	mV
Load Regulation LM309H LM309K	$T_j = 25^{\circ}$C $5\,mA \leq I_{OUT} \leq 0.5A$ $5\,mA \leq I_{OUT} \leq 1.5A$		20 50	50 100	mV mV
Output Voltage	$7V \leq V_{IN} \leq 25V$ $5\,mA \leq I_{OUT} \leq I_{max}$ $P < P_{max}$	4.75		5.25	V
Quiescent Current	$7V \leq V_{IN} \leq 25V$		5.2	10	mA
Quiescent Current Change	$7V \leq V_{IN} \leq 25V$ $5\,mA \leq I_{OUT} \leq I_{max}$			0.5 0.8	mA mA
Output Noise Voltage	$T_A = 25^{\circ}$C $10\,Hz \leq f \leq 100\,kHz$		40		μV
Long Term Stability				20	mV
Thermal Resistance Junction to Case (Note 2) LM309H LM309K			15 3.0		$^{\circ}$C/W $^{\circ}$C/W

Note 1: Unless otherwise specified, these specifications apply for $0^{\circ}C \leq T_j \leq 125^{\circ}C$, $V_{IN} = 10V$ and $I_{OUT} = 0.1A$ for the LM309H or $I_{OUT} = 0.5A$ for the LM309K. For the LM309H, $I_{max} = 0.2A$ and $P_{max} = 2.0W$. For the LM309K, $I_{max} = 1.0A$ and $P_{max} = 20W$.

Note 2: Without a heat sink, the thermal resistance of the TO-5 package is about 150°C/W, while that of the TO-3 package is approximately 35°C/W. With a heat sink, the effective thermal resistance can only approach the values specified, depending on the efficiency of the sink.

LM309

typical performance characteristics

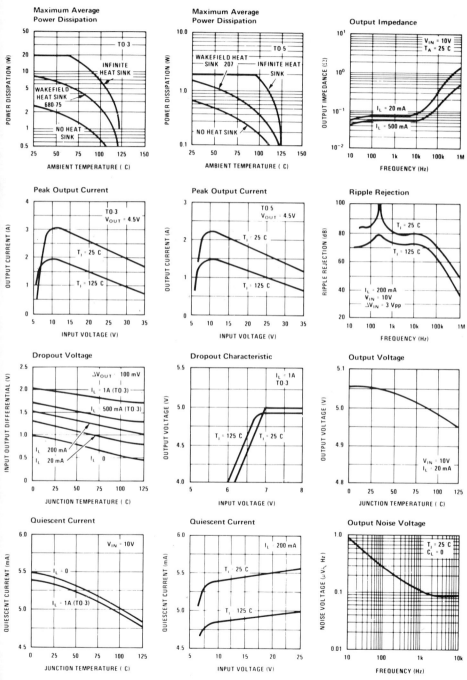

251

NATIONAL

LF155/LF156/LF157 monolithic JFET input operational amplifiers*

LF155, LF155A, LF255, LF355, LF355A low supply current
LF156, LF156A, LF256, LF356, LF356A wide band
LF157, LF157A, LF257, LF357, LF357A wide band decompensated ($A_{V_{MIN}}$ = 5)

general description

These are the first monolithic JFET input operational amplifiers to incorporate well matched, high voltage JFETs on the same chip with standard bipolar transistors (BI-FET Technology). These amplifiers feature low input bias and offset currents, low offset voltage and offset voltage drift, coupled with offset adjust which does not degrade drift or common-mode rejection. The devices are also designed for high slew rate, wide bandwidth, extremely fast settling time, low voltage and current noise and a low 1/f noise corner.

advantages

- Replace expensive hybrid and module FET op amps
- Rugged JFETs allow blow-out free handling compared with MOSFET input devices
- Excellent for low noise applications using either high or low source impedance—very low 1/f corner
- Offset adjust does not degrade drift or common-mode rejection as in most monolithic amplifiers
- New output stage allows use of large capacitive loads (10,000 pF) without stability problems
- Internal compensation and large differential input voltage capability

applications

- Precision high speed integrators
- Fast D/A and A/D converters
- High impedance buffers
- Wideband, low noise, low drift amplifiers

- Logarithmic amplifiers
- Photocell amplifiers
- Sample and Hold circuits

common features
(LF155A, LF156A, LF157A)

Low input bias current	30 pA
Low Input Offset Current	3 pA
High input impedance	$10^{12}\,\Omega$
Low input offset voltage	1 mV
Low input offset voltage temperature drift	$3\mu V/^\circ C$
Low input noise current	0.01 pA$/\sqrt{Hz}$
High common-mode rejection ratio	100 dB
Large dc voltage gain	106 dB

uncommon features

	LF155A	LF156A	LF157A (A_V = 5)	UNITS
Extremely fast settling time to 0.01%	4	1.5	1.5	μs
Fast slew rate	5	12	50	V/μs
Wide gain bandwidth	2.5	5	20	MHz
Low input noise voltage	20	12	12	nV/$\sqrt{Hz}$

BI-FET technology

simplified schematic

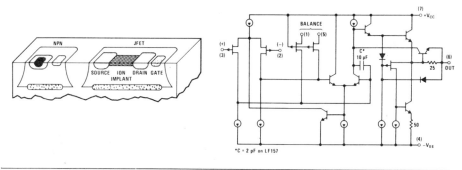

B15M95/Printed in U.S.A.

Courtesy of National Semiconductor Corporation

absolute maximum ratings

	LF155A/6A/7A	LF355A/6A/7A	LF155/6/7	LF255/6/7	LF355/6/7
Supply Voltage	±22V	±22V	±22V	±22V	±18V
Power Dissipation (Note 1) TO-99 (H package)	670 mW	500 mW	670 mW	570 mW	500 mW
Operating Temperature Range	-55°C to $+125^\circ$C	0°C to $+70^\circ$C	-55°C to $+125^\circ$C	-25°C to $+85^\circ$C	0°C to $+70^\circ$C
T_j(MAX)	150°C	100°C	150°C	110°C	100°C
Differential Input Voltage	±40V	±40V	±40V	±40V	±30V
Input Voltage Range (Note 2)	±20V	±20V	±20V	±20V	±16V
Output Short Circuit Duration	Continuous	Continuous	Continuous	Continuous	Continuous
Storage Temperature Range	-65°C to $+150^\circ$C	-65°C to $+150^\circ$C	-65°C to $+150^\circ$C	-65°C to $+150^\circ$C	-65°C to $+150^\circ$C
Lead Temperature (Soldering, 10 seconds)	300°C	300°C	300°C	300°C	300°C

dc electrical characteristics

(Note 3)

SYMBOL	PARAMETER	CONDITIONS	LF155A/6A/7A			LF355A/6A/7A			UNITS
			MIN	TYP	MAX	MIN	TYP	MAX	
V_{OS}	Input Offset Voltage	$R_S = 50\Omega$, $T_A = 25^\circ$C		1	2		1	2	mV
		Over Temperature			2.5			2.3	mV
$\Delta V_{OS}/\Delta T$	Average TC of Input Offset Voltage	$R_S = 50\Omega$		3	5		3	5	μV/$^\circ$C
$\Delta TC/\Delta V_{OS}$	Change in Average TC with V_{OS} Adjust	$R_S = 50\Omega$, (Note 4)		0.5			0.5		μV/$^\circ$C per mV
I_{OS}	Input Offset Current	$T_j = 25^\circ$C, (Notes 3, 5)		3	10		3	10	pA
		$T_j \leq T_{HIGH}$			10			1	nA
I_B	Input Bias Current	$T_J = 25^\circ$C, (Notes 3, 5)		30	50		30	50	pA
		$T_J \leq T_{HIGH}$			25			5	nA
R_{IN}	Input Resistance	$T_J = 25^\circ$C		10^{12}			10^{12}		Ω
A_{VOL}	Large Signal Voltage Gain	$V_S = \pm15$V, $T_A = 25^\circ$C $V_O = \pm10$V, $R_L = 2k$	50	200		50	200		V/mV
		Over Temperature	25			25			V/mV
V_O	Output Voltage Swing	$V_S = \pm15$V, $R_L = 10k$	±12	±13		±12	±13		V
V_{CM}	Input Common-Mode Voltage Range	$V_S = \pm15$V	±11	+15.1 −12		±11	+15.1 −12		V V
CMRR	Common-Mode Rejection Ratio		85	100		85	100		dB
PSRR	Supply Voltage Rejection Ratio	(Note 6)	85	100		85	100		dB

ac electrical characteristics

$T_A = 25^\circ$C, $V_S = \pm15$V

SYMBOL	PARAMETER	CONDITIONS	LF155A/355A			LF156A/356A			LF157A/357A			UNITS
			MIN	TYP	MAX	MIN	TYP	MAX	MIN	TYP	MAX	
SR	Slew Rate	LF155A/6A: $A_V = 1$, LF157A: $A_V = 5$	3	5		10	12		40	50		V/μs
GBW	Gain-Bandwidth Product			2.5		4	4.5		15	20		MHz
t_s	Settling Time to 0.01%	(Note 7)		4			1.5			1.5		μs
e_n	Equivalent Input Noise Voltage	$R_S = 100\Omega$ f = 100 Hz		25			15			15		nV/$\sqrt{\text{Hz}}$
		f = 1000 Hz		20			12			12		nV/$\sqrt{\text{Hz}}$
i_n	Equivalent Input Noise Current	f = 100 Hz		0.01			0.01			0.01		pA/$\sqrt{\text{Hz}}$
		f = 1000 Hz		0.01			0.01			0.01		pA/$\sqrt{\text{Hz}}$
C_{IN}	Input Capacitance			3			3			3		pF

dc electrical characteristics

(Note 3)

SYMBOL	PARAMETER	CONDITIONS	LF155/6/7			LF255/6/7			LF355/6/7			UNITS
			MIN	TYP	MAX	MIN	TYP	MAX	MIN	TYP	MAX	
V_{OS}	Input Offset Voltage	$R_S = 50\Omega$, $T_A = 25°C$		3	5		3	5		3	10	mV
		Over Temperature			7			6.5			13	mV
$\Delta V_{OS}/\Delta T$	Average TC of Input Offset Voltage	$R_S = 50\Omega$		5			5			5		$\mu V/°C$
$\Delta TC/\Delta V_{OS}$	Change in Average TC with V_{OS} Adjust	$R_S = 50\Omega$, (Note 4)		0.5			0.5			0.5		$\mu V/°C$ per mV
I_{OS}	Input Offset Current	$T_j = 25°C$, (Notes 3, 5)		3	20		3	20		3	50	pA
		$T_j \le T_{HIGH}$			20			1			2	nA
I_B	Input Bias Current	$T_J = 25°C$, (Notes 3, 5)		30	100		30	100		30	200	pA
		$T_J \le T_{HIGH}$			50			5			8	nA
R_{IN}	Input Resistance	$T_J = 25°C$		10^{12}			10^{12}			10^{12}		Ω
$A_{V_{OL}}$	Large Signal Voltage Gain	$V_S = \pm15V$, $T_A = 25°C$ $V_O = \pm10V$, $R_L = 2k$	50	200		50	200		25	200		V/mV
		Over Temperature	25			25			15			V/mV
V_O	Output Voltage Swing	$V_S = \pm15V$, $R_L = 10k$	±12	±13		±12	±13		±12	±13		V
V_{CM}	Input Common-Mode Voltage Range	$V_S = \pm15V$	±11	+15.1 −12		±11	+15.1 −12		±10	+15.1 −12		V V
CMRR	Common-Mode Rejection Ratio		85	100		85	100		80	100		dB
PSRR	Supply Voltage Rejection Ratio	(Note 6)	85	100		85	100		80	100		dB

dc electrical characteristics

$T_A = 25°C$, $V_S = \pm15V$

PARAMETER	LF155A/355A LF155/255		LF355		LF156A/356A LF156/256		LF356		LF157A/357A LF157/257		LF357		UNITS
	TYP	MAX	TYP	MAX	TYP	MAX	TYP	MAX	TYP	MAX	TYP	MAX	
Supply Current,	2	4	2	4	5	7	5	10	5	7	5	10	mA

ac electrical characteristics

$T_A = 25°C$, $V_S = \pm15V$

SYMBOL	PARAMETER	CONDITIONS	LF155/LF255/ LF355	LF156/LF256	LF156/LF256/ LF356	LF157/LF257	LF157/LF257/ LF357	UNITS
			TYP	MIN	TYP	MIN	TYP	
SR	Slew Rate	LF155/6: $A_V = 1$, LF157: $A_V = 5$	5	7.5	12	30	50	V/μs
GBW	Gain-Bandwidth Product		2.5		5		20	MHz
t_s	Settling Time to 0.01%	(Note 7)	4		1.5		1.5	μs
e_n	Equivalent Input Noise Voltage	$R_S = 100\Omega$ $f = 100$ Hz	25		15		15	nV/$\sqrt{Hz}$
		$f = 1000$ Hz	20		12		12	nV/$\sqrt{Hz}$
i_n	Equivalent Input Current Noise	$f = 100$ Hz	0.01		0.01		0.01	pA/$\sqrt{Hz}$
		$f = 1000$ Hz	0.01		0.01		0.01	pA/$\sqrt{Hz}$
C_{IN}	Input Capacitance		3		3		3	pF

notes for electrical characteristics

Note 1: The TO-99 package must be derated based on a thermal resistance of $150°C/W$ junction to ambient or $45°C/W$ junction to case.

Note 2: Unless otherwise specified the absolute maximum negative input voltage is equal to the negative power supply voltage.

Note 3: These specifications apply for $\pm15V \leq V_S \leq \pm20V$, $-55°C \leq T_A \leq +125°C$ and $T_{HIGH} = +125°C$ unless otherwise stated for the LF155A/6A/7A and the LF155/6/7. For the LF255/6/7, these specifications apply for $\pm15V \leq V_S \leq \pm20V$, $25°C \leq T_A \leq +85°C$ and $T_{HIGH} = 85°C$ unless otherwise stated. For the LF355A/6A/7A, these specifications apply for $\pm15V \leq V_S \leq \pm20V$, $0°C \leq T_A \leq +70°C$ and $T_{HIGH} = +70°C$, and for the LF355/6/7 these specifications apply for $V_S = \pm15V$ and $0°C \leq T_A \leq +70°C$. V_{OS}, I_B and I_{OS} are measured at $V_{CM} = 0$.

Note 4: The Temperature Coefficient of the adjusted input offset voltage changes only a small amount ($0.5\mu V/°C$ typically) for each mV of adjustment from its original unadjusted value. Common-mode rejection and open loop voltage gain are also unaffected by offset adjustment.

Note 5: The input bias currents are junction leakage currents which approximately double for every $10°C$ increase in the junction temperature, T_J. Due to limited production test time, the input bias currents measured are correlated to junction temperature. In normal operation the junction temperature rises above the ambient temperature as a result of internal power dissipation, Pd. $T_j = T_A + \Theta_{jA}$ Pd where Θ_{jA} is the thermal resistance from junction to ambient. Use of a heat sink is recommended if input bias current is to be kept to a minimum.

Note 6: Supply Voltage Rejection is measured for both supply magnitudes increasing or decreasing simultaneously, in accordance with common practice.

Note 7: Settling time is defined here, for a unity gain inverter connection using 2 kΩ resistors for the LF155/6. It is the time required for the error voltage (the voltage at the inverting input pin on the amplifier) to settle to within 0.01% of its final value from the time a 10V step input is applied to the inverter. For the LF157, $A_V = -5$, the feedback resistor from output to input is 2 kΩ and the output step is 10V (See Settling Time Test Circuit, page 9).

typical dc performance characteristics

Curves are for LF155, LF156 and LF157 unless otherwise specified.

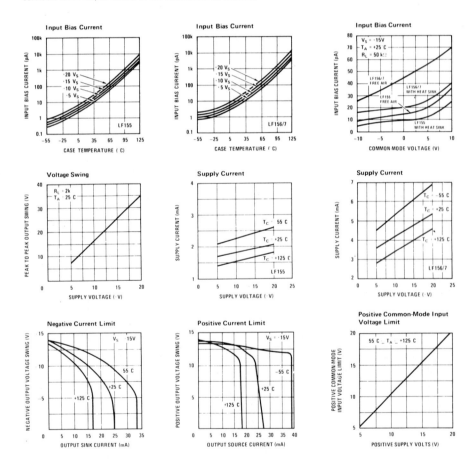

typical dc performance characteristics (con't)

Negative Common-Mode Input Voltage Limit

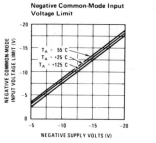

Open Loop Voltage Gain

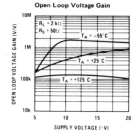

Output Voltage Swing

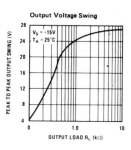

typical ac performance characteristics

Gain Bandwidth

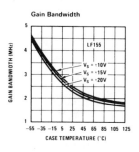

Unity Gain Bandwidth

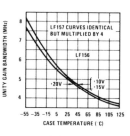

Normalized Slew Rate

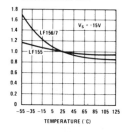

LF155 Small Signal Pulse Response, $A_V = +1$

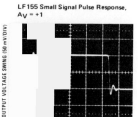

LF156 Small Signal Pulse Response, $A_V = +1$

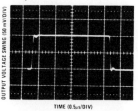

LF157 Small Signal Pulse Response, $A_V = +5$

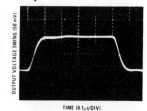

LF155 Large Signal Pulse Response, $A_V = +1$

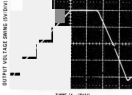

LF156 Large Signal Pulse Response, $A_V = +1$

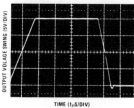

LF157 Large Signal Pulse Response, $A_V = +5$

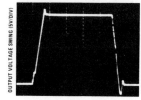

257

typical ac performance characteristics (con't)

Inverter Settling Time

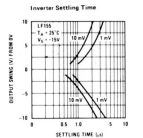

Inverter Settling Time

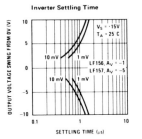

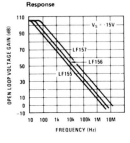

Bode Plot

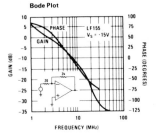

Bode Plot

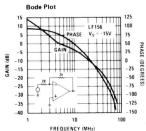

Bode Plot

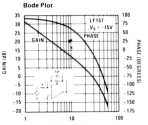

Common-Mode Rejection Ratio

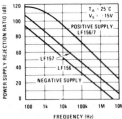

Power Supply Rejection Ratio

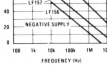

Power Supply Rejection Ratio

Undistorted Output Voltage Swing

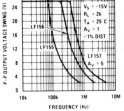

Equivalent Input Noise Voltage

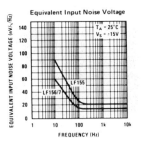

Equivalent Input Noise Voltage (Expanded Scale)

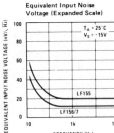

258

typical ac performance characteristics (con't)

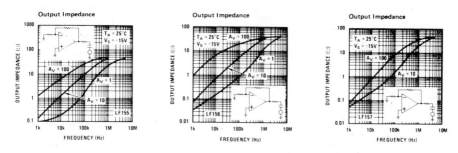

detailed schematic

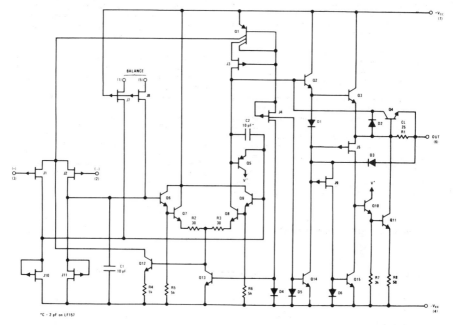

*C - 2 pF on LF157

connection diagram

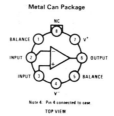

Metal Can Package

BALANCE (1) (8) NC
INPUT (2) (7) V⁺
INPUT (3) (6) OUTPUT
V⁻ (4) (5) BALANCE

Note 4: Pin 4 connected to case.
TOP VIEW

259

application hints

The LF155/6/7 series are op amps with JFET input devices. These JFETs have large reverse breakdown voltages from gate to source and drain eliminating the need for clamps across the inputs. Therefore large differential input voltages can easily be accomodated without a large increase in input current. The maximum differential input voltage is independent of the supply voltages. However, neither of the input voltages should be allowed to exceed the negative supply as this will cause large currents to flow which can result in a destroyed unit.

Exceeding the negative common-mode limit on either input will cause a reversal of the phase to the output and force the amplifier output to the corresponding high or low state. Exceeding the negative common-mode limit on both inputs will force the amplifier output to a high state. In neither case does a latch occur since raising the input back within the common-mode range again puts the input stage and thus the amplifier in a normal operating mode.

Exceeding the positive common-mode limit on a single input will not change the phase of the output however, if both inputs exceed the limit, the output of the amplifier will be forced to a high state.

These amplifiers will operate with the common-mode input voltage equal to the positive supply. In fact, the common-mode voltage can exceed the positive supply by approximately 100 mV independent of supply voltage and over the full operating temperature range. The positive supply can therefore be used as a reference on an input as, for example, in a supply current monitor and/or limiter.

Precautions should be taken to ensure that the power supply for the integrated circuit never becomes reversed

in polarity or that the unit is not inadvertently installed backwards in a socket as an unlimited current surge through the resulting forward diode within the IC could cause fusing of the internal conductors and result in a destroyed unit.

Because these amplifiers are JFET rather than MOSFET input op amps they do not require special handling.

All of the bias currents in these amplifiers are set by FET current sources. The drain currents for the amplifiers are therefore essentially independent of supply voltage.

As with most amplifiers, care should be taken with lead dress, component placement and supply decoupling in order to ensure stability. For example, resistors from the output to an input should be placed with the body close to the input to minimize "pickup" and maximize the frequency of the feedback pole by minimizing the capacitance from the input to ground.

A feedback pole is created when the feedback around any amplifier is resistive. The parallel resistance and capacitance from the input of the device (usually the inverting input) to ac ground set the frequency of the pole. In many instances the frequency of this pole is much greater than the expected 3 dB frequency of the closed loop gain and consequently there is negligible effect on stability margin. However, if the feedback pole is less than approximately six times the expected 3 dB frequency a lead capacitor should be placed from the output to the input of the op amp. The value of the added capacitor should be such that the RC time constant of this capacitor and the resistance it parallels is greater than or equal to the original feedback pole time constant.

typical circuit connections

V_{OS} Adjustment

- V_{OS} is adjusted with a 25k potentiometer
- The potentiometer wiper is connected to V^+
- For potentiometers with temperature coefficient of 100 ppm/°C or less the additional drift with adjust is ≈ 0.5μV/°C/mV of adjustment
- Typical overall drift: 5μV/°C · (0.5μV/°C/mV of adj.)

Driving Capacitive Loads

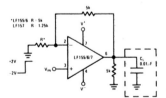

Due to a unique output stage design these amplifiers have the ability to drive large capacitive loads and still maintain stability. $C_{L\,MAX}$ · 0.01μF.
Overshoot ≤ 20%
Settling time (t_s) · 5μs

LF157. A Large Power BW Amplifier

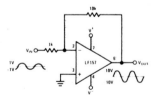

For distortion · 1% and a 20 Vp-p V_{OUT} swing, power bandwidth is: 500 kHz.

typical applications

Settling Time Test Circuit

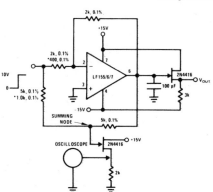

- Settling time is tested with the LF155/6 connected as unity gain inverter and LF157 connected for $A_V = -5$
- FET used to isolate the probe capacitance
- Output = 10V step

$*A_V = -5$ for LF157

Large Signal Inverter Output, V_{OUT} (from Settling Time Circuit)

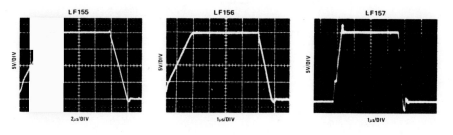

LF155	LF156	LF157
5V/DIV	5V/DIV	5V/DIV
2μs/DIV	1μs/DIV	1μs/DIV

Low Drift Adjustable Voltage Reference

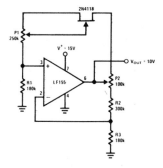

- $\Delta V_{OUT}/\Delta T = \pm 0.002\%/°C$
- All resistors and potentiometers should be wire-wound
- P1: drift adjust
- P2: V_{OUT} adjust
- Use LF155 for
  - ▲ Low I_B
  - ▲ Low drift
  - ▲ Low supply current

typical applications (con't)

Fast Logarithmic Converter

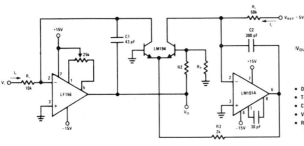

$$|V_{OUT}| = \left[1 + \frac{R2}{R_T}\right] \frac{kT}{q} \ln V_i \left[\frac{R_i}{V_{REF} \, R_i}\right] = \log V_i \frac{1}{R_i I_i}$$

R2 = 15.7k, R_T = 1k, 0.3%/°C (for temperature compensation)

- Dynamic range: 100μA $\leq$ I$_i$ $\leq$ 1 mA (5 decades), |V$_O$| = 1V/decade
- Transient response: 3μs for ΔI$_i$ = 1 decade
- C1, C2, R2, R3: added dynamic compensation
- V$_{OS}$ adjust the LF156 to minimize quiescent error
- R$_T$: Tel Labs type Q81 + 0.3%/°C.

Precision Current Monitor

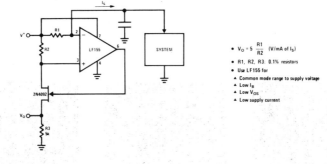

- $V_O = 5 \frac{R1}{R2}$ (V/mA of I$_S$)
- R1, R2, R3: 0.1% resistors
- Use LF155 for
  ▲ Common mode range to supply voltage
  ▲ Low I$_B$
  ▲ Low V$_{OS}$
  ▲ Low supply current

LF156 as an Output Amplifier in a Fast 8-Bit DAC

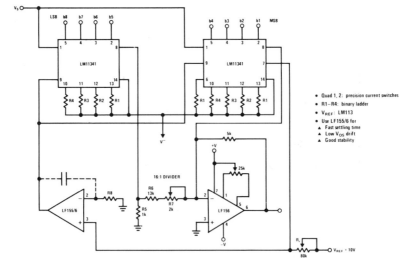

- Quad 1, 2: precision current switches
- R1–R4: binary ladder
- V$_{REF}$: LM113
- Use LF155/6 for
  ▲ Fast settling time
  ▲ Low V$_{OS}$ drift
  ▲ Good stability

262

typical applications (con't)

Wide BW Low Noise, Low Drift Amplifier

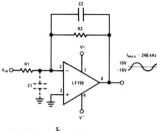

- Power BW: $f_{MAX} = \dfrac{S_r}{2\pi V_p} \cong 240\ kHz$

- Parasitic input capacitance C1 $\cong$ (3 pF for LF155, LF156, and LF157 plus any additional layout capacitance) interacts with feedback elements and creates undesirable high frequency pole. To compensate add C2 such that: R2C2 $\cong$ R1C1.

Isolating Large Capacitive Loads

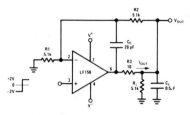

- Overshoot 6%
- t_s 10μs
- When driving large C_L the V_{OUT} slew rate determined by C_L and $I_{OUT\ MAX}$:

$$\frac{\Delta V_{OUT}}{\Delta T} = \frac{I_{OUT}}{C_L} \cong \frac{0.02}{0.5}\quad V/\mu s = 0.04\ V/\mu s$$

(with C_L shown)

Boosting the LF156 with a Current Amplifier

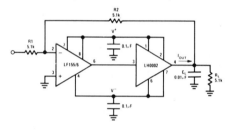

- $I_{OUT\ MAX} \ge 150\ mA$ (will drive $R_L \ge 100\Omega$)
- $\dfrac{\Delta V_{OUT}}{\Delta T} = \dfrac{0.15}{10^{-2}}$ V/μs = 15 V/μs (with C_L shown)
- No additional phase shift added by the current amplifier

Low Drift Peak Detector

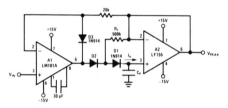

- By adding D1 and R_f, V_{D1} = 0 during hold mode. Leakage of D2 provided by feedback path through R_f.
- Leakage of circuit is essentially I_b (LF155, LF156) plus capacitor leakage of C_p.
- Diode D3 clamps V_{OUT} (A1) to V_{IN} V_{D3} to improve speed and to limit reverse bias of D2.
- Maximum input frequency should be $\ll 1/2\pi R_f C_{D2}$ where C_{D2} is the shunt capacitance of D2.

3 Decades VCO

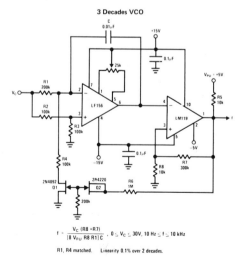

$$f = \frac{V_C\ (R8 + R7)}{|8\ V_{FU}\ R8\ R1|\ C}\ ,\ 0 \le V_C \le 30V,\ 10\ Hz \le f \le 10\ kHz$$

R1, R4 matched.   Linearity 0.1% over 2 decades.

263

typical applications (con't)

High Impedance, Low Drift Instrumentation Amplifier

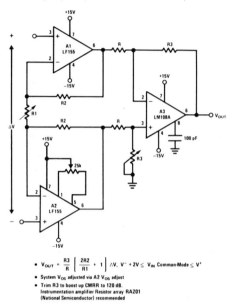

- $V_{OUT} = \dfrac{R3}{R} \left[\dfrac{2R2}{R1} + 1 \right] \Delta V, \; V^- + 2V \le V_{IN}$ Common-Mode $\le V^+$
- System V_{OS} adjusted via A2 V_{OS} adjust
- Trim R3 to boost up CMRR to 120 dB. Instrumentation amplifier Resistor array RA201 (National Semiconductor) recommended

Fast Sample and Hold

- Both amplifiers (A1, A2) have feedback loops individually closed with stable responses (overshoot negligible)
- Acquisition time, T_A, estimated by:

$$T_A \ge \left[\dfrac{2R_{ON} \cdot V_{IN} \cdot C_h}{S_r} \right]^{1/2} \; \text{provided that:}$$

$$V_{IN} < 2 \bullet S, \; R_{ON} \, C_h \; \text{and} \; T_A > \dfrac{V_{IN} \, C_h}{I_{OUT \, MAX}} \; , \; R_{ON} \text{ is of SW1}$$

If inequality not satisfied: $T_A > \dfrac{V_{IN} \, C_h}{20 \text{ mA}}$

- LF156 develops full S, output capability for $V_{IN} \ge 1V$
- Addition of SW2 improves accuracy by putting the voltage drop across SW1 inside the feedback loop
- Overall accuracy of system determined by the accuracy of both amplifiers, A1 and A2

264

typical applications (con't)

High Accuracy Sample and Hold

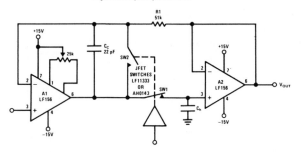

- By closing the loop through A2 the V_{OUT} accuracy will be determined uniquely by A1. No V_{OS} adjust required for A2.
- T_A can be estimated by same considerations as previously but, because of the added propagation delay in the feedback loop (A2) the overshoot is not negligible.
- Overall system slower than fast sample and hold
- R1, C_C: additional compensation
- Use LF156 for
  ▲ Fast settling time
  ▲ Low V_{OS}

High Q Band Pass Filter

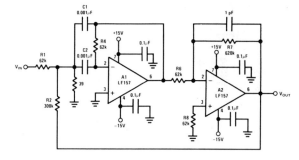

- By adding positive feedback (R2) Q increases to 40
- f_{BP} = 100 kHz
- $\dfrac{V_{OUT}}{V_{IN}} = 10\sqrt{Q}$
- Clean layout recommended
- Response to a 1 Vp-p tone burst: 300μs

High Q Notch Filter

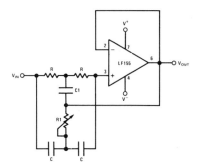

- 2R1 = R = 10 MΩ
  2C = C1 = 300 pF
- Capacitors should be matched to obtain high Q
- f_{NOTCH} = 120 Hz, notch = -55 dB, Q > 100
- Use LF155 for
  ▲ Low I_B
  ▲ Low supply current

definition of terms

Input Offset Voltage: That voltage which must be applied between the input terminals through two equal resistances to obtain zero output voltage.

Input Offset Current: The difference in the currents into the two input terminals when the output is at zero.

Input Bias Current: The average of the two input currents.

Input Common-Mode Voltage Range: The range of voltages on the input terminals for which the amplifier is operational. Note that the specifications are not guaranteed over the full common-mode voltage range unless specifically stated.

Common-Mode Rejection Ratio: The ratio of the input common-mode voltage range to the peak-to-peak change in input offset voltage over this range.

Input Resistance: The ratio of the change in input voltage to the change in input current on either input with the other grounded.

Supply Current: The current required from the power supply to operate the amplifier with no load and the output midway between the supplies.

Output Voltage Swing: The peak output voltage swing, referred to zero, that can be obtained without clipping.

Large-Signal Voltage Gain: The ratio of the output voltage swing to the change in input voltage required to drive the output from zero to this voltage.

Power Supply Rejection Ratio: The ratio of the change in input offset voltage to the change in power supply voltage producing it. The typical curves in this sheet show values for each supply independently changed. The electrical specification, however, is measured for both supply magnitudes increasing or decreasing simultaneously, in accordance with common practice.

Settling Time: The time required for the error between input and output to settle to within a specified limit after an input is applied to the test circuit shown in typical applications.

physical dimensions

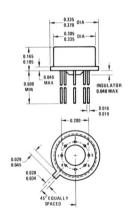

TO-99 Metal Can Package (H)
Order Number:

LF155AH	LF157H	LF356AH
LF156AH	LF255H	LF357AH
LF157AH	LF256H	LF355H
LF155H	LF257H	LF356H
LF156H	LF355AH	LF357H

Manufactured under one or more of the following U.S. patents: 3083262, 3189758, 3231797, 3303356, 3317671, 3323071, 3381071, 3408542, 3421025, 3426423, 3440498, 3518750, 3519897, 3557431, 3560765, 3566218, 3571630, 3575609, 3579059, 3593069, 3597640, 3607469, 3617859, 3631312, 3633052, 3638131, 3648071, 3651565, 3693248.

National Semiconductor Corporation
2900 Semiconductor Drive, Santa Clara, California 95051, (408) 732-5000/TWX (910) 339-9240
National Semiconductor GmbH
808 Fuerstenfeldbruck, Industriestrasse 10, West Germany, Tele. (08141) 1371/Telex 05-27649
National Semiconductor (UK) Ltd.
Larkfield Industrial Estate, Greenock, Scotland, Tele. (0475) 33251/Telex 778-632

ℝℂ/ℐ

**Solid State
Division**

Linear Integrated Circuits

Monolithic Silicon

**CA3140BT CA3140AT CA3140T
CA3140BS CA3140AS CA3140S**

File Number **956**

8-Lead TO-5
With Dual-In-Line
Formed Leads
"DIL-CAN"
S Suffix

8-Lead TO-5
T Suffix

H-1787   H-1528

BiMOS Operational Amplifiers*

With MOS/FET Input, Bipolar Output

Features:

- MOS/FET Input Stage
  - (a) Very high input impedance (Z_{IN}) — 1.5 TΩ typ.
  - (b) Very low input current (I_I) — 10 pA typ. at ± 15 V
  - (c) Low input-offset voltage (V_{IO}) — to 2 mV max.
  - (d) Wide common-mode input-voltage range (V_{ICR}) — can be swung 0.5 volt below negative supply-voltage rail
  - (e) Output swing complements input common-mode range
  - (f) Rugged input stage — bipolar diode protected
- Directly replaces industry type 741 in most applications

The CA3140B, CA3140A, and CA3140 are integrated-circuit operational amplifiers that combine the advantages of high-voltage PMOS transistors with high-voltage bipolar transistors on a single monolithic chip. Because of this unique combination of technologies, this device can now provide designers, for the first time, with the special performance features of the CA3130 COS/MOS operational amplifiers and the versatility of the 741 series of industry-standard operational amplifiers.

The CA3140, CA3140A, and CA3140 BiMOS operational amplifiers feature gate-protected MOS/FET (PMOS) transistors in the input circuit to provide very-high-input impedance, very-low-input current, and high-speed performance. The CA3140B operates at supply voltages from 4 to 44 volts; the CA3140A and CA3140 from 4 to 36 volts (either single or dual supply). These operational amplifiers are internally phase-compensated to achieve stable operation in unity-gain follower operation, and, additionally, have access terminals for a supplementary external capacitor if additional frequency roll-off is desired. Terminals are also provided for use in applications requiring input offset-voltage nulling. The use of PMOS field-effect transistors in the input stage results in common-mode input-voltage capability down to 0.5 volt below the negative-supply terminal, an important attribute for single-supply applications. The output stage uses bipolar transistors and includes built-in protection against damage from load-terminal short-circuiting to either supply-rail or to ground.

The CA3140 Series has the same 8-lead terminal pin-out used for the "741" and other

- Includes numerous industry operational amplifier categories such as general-purpose, FET input, wideband (high slew rate)
- Operation from 4-to-44 volts Single or Dual supplies
- Internally compensated
- Characterized for ± 15-volt operation and for TTL supply systems with operation down to 4 volts
- Wide bandwidth — 4.5 MHz unity gain at ± 15 V or 30 V; 3.7 MHz at 5 V
- High voltage-follower slew rate — 9 V/μs
- Fast settling time — 1.4 μs typ. to 10 mV with a 10-V_{p-p} signal
- Output swings to within 0.2 volt of negative supply
- Strobable output stage

Applications:

- Ground-referenced single-supply amplifiers in automobile and portable instrumentation
- Sample and hold amplifiers
- Long-duration timers/multivibrators (microseconds—minutes—hours)
- Photocurrent instrumentation
- Peak detectors   ■ Active filters
- Comparators
- Interface in 5 V TTL systems & other low-supply voltage systems
- All standard operational amplifier applications
- Function generators   ■ Tone controls
- Power supplies   ■ Portable instruments
- Intrusion alarm systems

CA3140B, CA3140A, CA3140 BiMOS Operational Amplifiers

Courtesy of RCA Solid State Division

267

industry-standard operational amplifiers. They are supplied in either the standard 8-lead TO-5 style package (T suffix), or in the 8-lead dual-in-line formed-lead TO-5 style package "DIL-CAN" (S suffix). The CA3140B is intended for operation at supply voltages ranging from 4 to 44 volts, for applications requiring premium-grade specifications and with electrical limits established for operation over the range from −55°C to + 125°C. The CA3140A and CA3140 are for operation at supply voltages up to 36 volts (±18 volts). The CA3140 and CA3140A can also be operated safely over the temperature range from −55°C to + 125°C, although specification limits for their electrical parameters do not apply when they are operated beyond their specified temperature ranges.

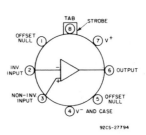

Fig.1 − Functional diagram of CA3140 series.

TYPICAL ELECTRICAL CHARACTERISTICS

CHARACTERISTIC	TEST CONDITIONS V^+ = +15 V V^- = −15 V T_A = 25°C	LIMITS			UNITS
		CA3140B	CA3140A	CA3140	
Input Offset Voltage Adjustment Resistor	Typ. Value of Resistor Between Term. 4 and 5 or 4 and 1 to Adjust Max. V_{IO}	43	18	4.7	kΩ
Input Resistance R_1		1.5	1.5	1.5	TΩ
Input Capacitance C_I		4	4	4	pF
Output Resistance R_O		60	60	60	Ω
Equivalent Wideband Input Noise Voltage e_n (See Fig.41)	BW = 140 kHz R_S = 1 MΩ	48	48	48	μV
Equivalent Input Noise Voltage e_n (See Fig.10)	f = 1 kHz R_S = 100 Ω	40	40	40	nV/√Hz
	f = 10 kHz	12	12	12	
Short-Circuit Current to Opposite Supply Source I_{OM}^+		40	40	40	mA
Sink I_{OM}^-		18	18	18	mA
Gain-Bandwidth Product, (See Figs. 5 &18) f_T		4.5	4.5	4.5	MHz
Slew Rate, (See Fig.6) SR		9	9	9	V/μs
Sink Current From Terminal 8 To Terminal 4 to Swing Output Low		220	220	220	μA
Transient Response: Rise Time t_r Overshoot (See Fig. 40)	R_L = 2 kΩ C_L = 100 pF	0.08	0.08	0.08	μs
		10	10	10	%
Settling Time at 10 V_{p-p}, (See Fig.17) t_s 1 mV	R_L = 2 kΩ C_L = 100 pF Voltage Follower	4.5	4.5	4.5	μs
10 mV		1.4	1.4	1.4	

ELECTRICAL CHARACTERISTICS FOR EQUIPMENT DESIGN

At V^+ = 15 V, V^- = 15 V, T_A = 25°C Unless Otherwise Specified

CHARACTERISTIC	LIMITS									UNITS
	CA3140B			CA3140A			CA3140			
	Min.	Typ.	Max.	Min.	Typ.	Max.	Min.	Typ.	Max.	
Input Offset Voltage, $|V_{IO}|$	–	0.8	2	–	2	5	–	5	15	mV
Input Offset Current, $|I_{IO}|$	–	0.5	10	–	0.5	20	–	0.5	30	pA
Input Current, I_I	–	10	30	–	10	40	–	10	50	pA
Large-Signal Voltage Gain, A_{OL}● (See Figs. 4,18)	50 k	100 k	–	20 k	100 k	–	20 k	100 k	–	V/V
	94	100	–	86	100	–	86	100	–	dB
Common-Mode Rejection Ratio, CMRR (See Fig.9)	–	20	50	–	32	320	–	32	320	μV/V
	86	94	–	70	90	–	70	90	–	dB
Common-Mode Input-Voltage Range, V_{ICR} (See Fig.20)	–15	–15.5 to +12.5	12	–15	–15.5 to +12.5	12	–15	–15.5 to +12.5	11	V
Power-Supply Rejection Ratio, PSRR $\overline{\Delta V_{IO}/\Delta V}$ (See Fig.11)	–	32	100	–	100	150	–	100	150	μV/V
	80	90	–	76	80	–	76	80	–	dB
Max. Output Voltage■ V_{OM}^+ (See Figs.13,20) $\overline{V_{OM}^-}$	+12	13	–	+12	13	–	+12	13	–	V
	–14	–14.4	–	–14	–14.4	–	–14	–14.4	–	
Supply Current, I^+ (See Fig.7)	–	4	6	–	4	6	–	4	6	mA
Device Dissipation, P_D	–	120	180	–	120	180	–	120	180	mW
Input Current, I_I▲ (See Fig.19)	–	10	30	–	10	–	–	10	–	nA
Input Offset Voltage V_{IO}▲	–	1.3	3	–	3	–	–	10	–	mV
Large-Signal Voltage Gain, A_{OL}▲ (See Figs.4,18)	20 k	100 k	–	–	100 k	–	–	100 k	–	V/V
	86	100	–	–	100	–	–	100	–	dB
Max. Output Voltage,★ V_{OM}^+ $\overline{V_{OM}^-}$	+19	+19.5	–	–	–	–	–	–	–	V
	–21	–21.4	–	–	–	–	–	–	–	
Large-Signal Voltage Gain, A_{OL}♦★	20 k	50 k	–	–	–	–	–	–	–	V/V
	86	94	–	–	–	–	–	–	–	dB

● At V_O = 26$V_{p\text{-}p}$, +12V, –14V and R_L = 2 kΩ.

■ At R_L = 2 kΩ.

▲ At T_A = –55°C to +125°C, V^+ = 15 V, V^- = 15 V, V_O = 26$V_{p\text{-}p}$, R_L = 2 kΩ.

★ At V^+ = 22 V, V^- = 22 V.

♦ At V_O = +19 V, –21 V, and R_L = 2 kΩ.

MAXIMUM RATINGS, *Absolute-Maximum Values:*

	CA3140, CA3140A	CA3140B

DC SUPPLY VOLTAGE

(BETWEEN V$^+$ AND V$^-$ TERMINALS) . . . . 36 V 44 V

DIFFERENTIAL-MODE INPUT VOLTAGE . . . . $\pm$8 V $\pm$8 V

COMMON-MODE DC INPUT VOLTAGE . . . . (V$^+$ +8 V) to (V$^-$ −0.5 V)

INPUT-TERMINAL CURRENT . . . . 1 mA

DEVICE DISSIPATION:

  WITHOUT HEAT SINK −

    UP TO 55°C. . . . . 630 mW

    ABOVE 55°C . . . . Derate linearly 6.67 mW/^{o}C

  WITH HEAT SINK −

    Up to 55°C. . . . . 1 W

    Above 55°C . . . . Derate linearly 16.7 mW/^{o}C

TEMPERATURE RANGE:

  OPERATING . . . . −55 to + 125°C

  STORAGE . . . . −65 to + 150°C

OUTPUT SHORT-CIRCUIT DURATION* . . . . INDEFINITE

LEAD TEMPERATURE (DURING SOLDERING): .

  AT DISTANCE 1/16 $\pm$ 1/32 INCH (1.59 $\pm$ 0.79 MM)

  FROM CASE FOR 10 SECONDS MAX. . . . . +265°C

* Short circuit may be applied to ground or to either supply.

TYPICAL ELECTRICAL CHARACTERISTICS FOR DESIGN GUIDANCE

At V$^+$ = 5 V, V$^-$ = 0 V, T$_A$ = 25°C

CHARACTERISTIC		LIMITS			UNITS
		CA3140B	CA3140A	CA3140	
Input Offset Voltage	|V$_{IO}$|	0.8	2	5	mV
Input Offset Current	|I$_{IO}$|	0.1	0.1	0.1	pA
Input Current	I$_I$	2	2	2	pA
Input Resistance		1	1	1	TΩ
Large-Signal Voltage Gain	A$_{OL}$	100 k	100 k	100 k	V/V
(See Figs.4,18)		100	100	100	dB
Common-Mode Rejection Ratio,	CMRR	20	32	32	μV/V
		94	90	90	dB
Common-Mode Input-Voltage Range	V$_{ICR}$	−0.5	−0.5	−0.5	V
(See Fig.20)		2.6	2.6	2.6	
Power-Supply Rejection Ratio	ΔV$_{IO}$/ΔV$^+$	32	100	100	μV/V
		90	80	80	dB
Maximum Output Voltage	V$_{OM}$$^+$	3	3	3	V
(See Figs.13,20)	$\overline{V_{OM}}^-$	0. 13	0.13	0.13	
Maximum Output Current:					
Source	I$_{OM}$$^+$	10	10	10	mA
Sink	I$_{OM}$$^-$	1	1	1	
Slew Rate (See Fig.6)		7	7	7	V/μs
Gain-Bandwidth Product (See Fig.5)	f$_T$	3.7	3.7	3.7	MHz
Supply Current (See Fig.7)	I$^+$	1.6	1.6	1.6	mA
Device Dissipation	P$_D$	8	8	8	mW
Sink Current from Term. 8 to Term. 4 to Swing Output Low		200	200	200	μA

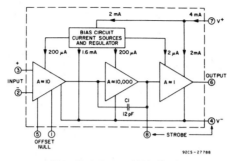

Fig.2 – Block diagram of CA3140 series.

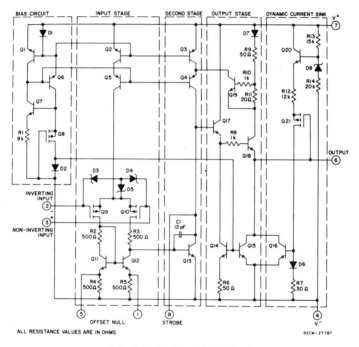

Fig.3 – Schematic diagram of CA3140 series.

CIRCUIT DESCRIPTION

Fig.2 is a block diagram of the CA3140 Series PMOS Operational Amplifiers. The input terminals may be operated down to 0.5 V below the negative supply rail. Two class A amplifier stages provide the voltage gain, and a unique class AB amplifier stage provides the current gain necessary to drive low-impedance loads.

A biasing circuit provides control of cascoded constant-current flow circuits in the first and second stages. The CA3140 includes an on-

chip phase-compensating capacitor that is sufficient for the unity gain voltage-follower configuration.

Input Stages – The schematic circuit diagram of the CA3140 is shown in Fig.3. It consists of a differential-input stage using PMOS field-effect transistors (Q9, Q10) working into a mirror pair of bipolar transistors (Q11, Q12) functioning as load resistors together with resistors R2 through R5. The mirror-pair transistors also function as a differen-

271

tial-to-single-ended converter to provide base-current drive to the second-stage bipolar transistor (Q13). Offset nulling, when desired, can be effected with a 10-kΩ potentiometer connected across terminals 1 and 5 and with its slider arm connected to terminal 4. Cascode-connected bipolar transistors Q2, Q5 are the constant-current source for the input stage. The base-biasing circuit for the constant-current source is described subsequently. The small diodes D3, D4, D5 provide gate-oxide protection against high-voltage transients, e.g., static electricity.

Second Stage — Most of the voltage gain in the CA3140 is provided by the second amplifier stage, consisting of bipolar transistor Q13 and its cascode-connected load resistance provided by bipolar transistors Q3, Q4. On-chip phase compensation, sufficient for a majority of the applications is provided by C1. Additional Miller-Effect compensation (roll-off) can be accomplished, when desired, by simply connecting a small capacitor between terminals 1 and 8. Terminal 8 is also used to strobe the output stage into quiescence. When terminal 8 is tied to the negative supply rail (terminal 4) by mechanical or electrical means, the output terminal 6 swings low, i.e., approximately to terminal 4 potential.

Output Stage — The CA3140 Series circuits employ a broadband output stage that can sink loads to the negative supply to complement the capability of the PMOS input stage when operating near the negative rail. Quiescent current in the emitter-follower cascade circuit (Q17, Q18) is established by transistors (Q14, Q15) whose base-currents are "mirrored" to current flowing through diode D2 in the bias circuit section. When the CA3140 is operating such that output terminal 6 is sourcing current, transistor Q18 functions as an emitter-follower to source current from the V+ bus (terminal 7), via D7, R9, and R11. Under these conditions, the collector potential of Q13 is sufficiently high to permit the necessary flow of base current to emitter follower Q17 which, in turn, drives Q18.

When the CA3140 is operating such that output terminal 6 is sinking current to the V— bus, transistor Q16 is the current-sinking element. Transistor Q16 is mirror-connected to D6, R7, with current fed by way of Q21, R12, and Q20. Transistor Q20, in turn, is biased by current-flow through R13, zener D8, and R14. The dynamic current-sink is controlled by voltage-level sensing. For purposes of explanation, it is assumed that output terminal 6 is quiescently established at the potential mid-point between the V+ and V— supply rails. When output-current sinking-mode operation is required, the collector potential of transistor Q13 is driven below its quiescent level, thereby causing Q17, Q18 to decrease the output voltage at terminal 6. Thus, the gate terminal of PMOS transistor Q21 is displaced toward the V— bus, thereby reducing the channel resistance of Q21. As a consequence, there is an incremental increase in current flow through Q20, R12, Q21, D6, R7, and the base of Q16. As a result, Q16 sinks current from terminal 6 in direct response to the incremental change in output voltage caused by Q18. This sink current flows regardless of load; any excess current is internally supplied by the emitter-follower Q18. Short-circuit protection of the output circuit is provided by Q19, which is driven into conduction by the high voltage drop developed across R11 under output short-circuit conditions. Under these conditions, the collector of Q19 diverts current from Q4 so as to reduce the base-current drive from Q17, thereby limiting current flow in Q18 to the short-circuited load terminal.

Bias Circuit — Quiescent current in all stages (except the dynamic current sink) of the CA3140 is dependent upon bias current flow in R1. The function of the bias circuit is to establish and maintain constant-current flow through D1, Q6, Q8 and D2. D1 is a diode-connected transistor mirror-connected in parallel with the base-emitter junctions of Q1, Q2, and Q3. D1 may be considered as a current-sampling diode that senses the emitter current of Q6 and automatically adjusts the base current of Q6 (via Q1) to maintain a constant current through Q6, Q8, D2. The base-currents in Q2, Q3 are also determined by constant-current flow D1. Furthermore, current in diode-connected transistor D2 establishes the currents in transistors Q14 and Q15.

HANDLING AND OPERATING CONSIDERATIONS

The CA3140 uses MOS field-effect transistors in the input circuit. Although the CA3140 utilizes rugged bipolar diodes for protection of the input circuit it is good practice in the handling, testing, and operation of these devices to use the following recommended procedures:

1. Soldering-iron tips, metal parts of fixtures, tools, and handling facilities should be grounded.
2. Devices should not be inserted into or removed from circuits with the power ON because transient voltages may cause damage.

3. Signals should not be applied to the input (Terms. 2 and 3) when the device power supply is OFF. Input-terminal currents should not exceed 1 mA.

4. After these devices have been mounted on circuit boards, proper handling precautions should still be observed if the input terminals are unterminated. It is good practice during board-processing operations to return Terms. 2 and 3 to Term. 4 by jumping the appropriate conductors.

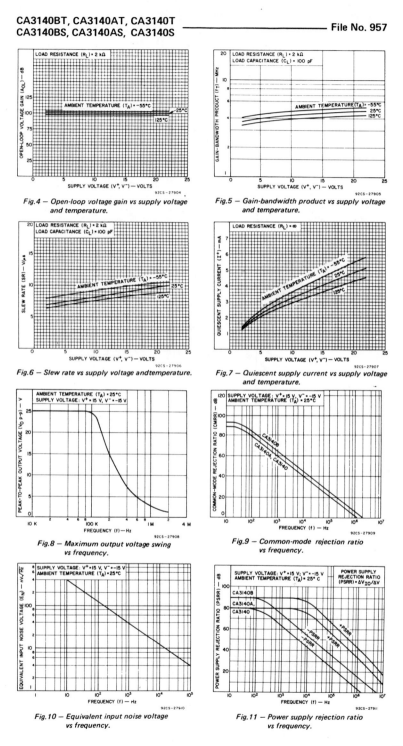

Fig.4 — Open-loop voltage gain vs supply voltage and temperature.

Fig.5 — Gain-bandwidth product vs supply voltage and temperature.

Fig.6 — Slew rate vs supply voltage and temperature.

Fig.7 — Quiescent supply current vs supply voltage and temperature.

Fig.8 — Maximum output voltage swing vs frequency.

Fig.9 — Common-mode rejection ratio vs frequency.

Fig.10 — Equivalent input noise voltage vs frequency.

Fig.11 — Power supply rejection ratio vs frequency.

273

Intersil

8038

PRECISION WAVEFORM GENERATOR/VOLTAGE CONTROLLED OSCILLATOR*

FEATURES

- Low Frequency Drift With Temperature — 50ppm/°C Max.
- Simultaneous Outputs — Sine-Wave, Square-Wave and Triangle.
- High Level Outputs — T^2L to 28V
- Low Distortion — 1%
- High Linearity — 0.1%
- Easy to Use — 50% Reduction in External Components.
- Wide Frequency Range of Operation 0.001Hz to 1.0MHz
- Variable Duty Cycle — 2% to 98%

GENERAL DESCRIPTION

The 8038 Waveform Generator is a monolithic integrated circuit, capable of producing sine, square, triangular, sawtooth and pulse waveform of high accuracy with a minimum of external components (refer to Figures 8 and 9) The frequency (or repetition rate) can be selected externally over a range from less than 1/1000 Hz to more than 1MHz and is highly stable over a wide temperature and supply voltage range. Frequency modulation and sweeping can be accomplished with an external voltage and the frequency can be programmed digitally through the use of either resistors or capacitors. The Waveform Generator utilizes advanced monolithic technology, such as thin film resistors and Schottky-barrier diodes. The 8038 Voltage Controlled Oscillator can be interfaced with phase lock loop circuitry to reduce temperature drift to below 50ppm/°C.

Courtesy of Intersil, Inc.

CONNECTION DIAGRAM

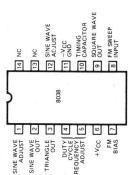

Pin	Label
1	SINE WAVE ADJUST
2	SINE WAVE OUT
3	TRIANGLE OUT
4	DUTY CYCLE / FREQUENCY ADJUST
5	DUTY CYCLE / FREQUENCY ADJUST
6	+VCC
7	FM BIAS
8	FM SWEEP INPUT
9	SQUARE WAVE OUT
10	TIMING CAPACITOR
11	−VCC GND
12	SINE WAVE ADJUST
13	NC
14	NC

8038

FUNCTIONAL DIAGRAM

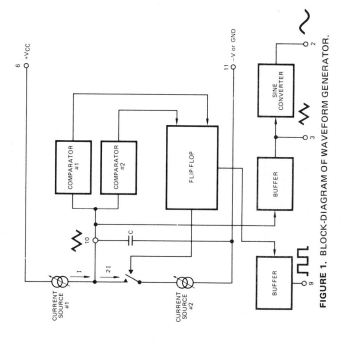

FIGURE 1. BLOCK-DIAGRAM OF WAVEFORM GENERATOR.

INTERSIL, INC., 10900 N. TANTAU AVE., CUPERTINO, CA 95014   (408) 996-5000 TWX 910-338-0228

Printed in U.S.A.

MAXIMUM RATINGS

Supply Voltage	±18V or 36V Total
Power Dissipation	750mW (Note 5)
Input Voltage (any pin)	Not To Exceed Supply Voltages
Input Current (Pins 4 and 5)	25mA
Output Sink Current (Pins 3 and 9)	25mA
Storage Temperature Range	-65°C to +125°C
Operating Temperature Range:	
8038AM, 8038BM	-55°C to +125°C
8038AC, 8038BC, 8038CC	0°C to +70°C

ELECTRICAL CHARACTERISTICS

$(V_S = \pm10V$ or $+20V$, $T_A = 25°C$, $R_L = 10\ K\Omega$ Unless Otherwise Specified) Note 3.

GENERAL CHARACTERISTICS	8038CC MIN	8038CC TYP	8038CC MAX	8038BC/BM MIN	8038BC/BM TYP	8038BC/BM MAX	8038AC/AM MIN	8038AC/AM TYP	8038AC/AM MAX	UNITS
Supply Voltage Operating Range										
Single Supply	+10		+30	+10		30	+10		30	V
Dual Supplies	±5		±15	±5		±15	±5		±15	V
Supply Current ($V_S = \pm10V$) Note 1.										
8038AM, 8038BM					12	15		12	15	mA
8038AC, 8038BC, 8038CC		12	20		12	20		12	20	mA

FREQUENCY CHARACTERISTICS (all waveforms)

	8038CC MIN	8038CC TYP	8038CC MAX	8038BC/BM MIN	8038BC/BM TYP	8038BC/BM MAX	8038AC/AM MIN	8038AC/AM TYP	8038AC/AM MAX	UNITS
Maximum Frequency of Oscillation	100,000			100,000			100,000			Hz
Sweep Frequency of FM		10			10			10		kHz
Sweep FM Range (Note 2)		40:1			40:1			40:1		
FM Linearity 10:1 Ratio		0.5			0.2			0.2		%
Frequency Drift With Temperature Note 6		50			50	100		20	50	ppm/°C
Frequency Drift With Supply Voltage (Over Supply Voltage Range)		0.05			0.05			0.05		%/Vs
Recommended Programming Resistors (R_A and R_B)	1000		1M	1000		1M	1000		1M	Ω

OUTPUT CHARACTERISTICS

Parameter	Min	Typ	Max	Min	Typ	Max	Min	Typ	Max	Unit
Square-Wave										
Leakage Current (Vq = 30v)			1			1			1	μA
Saturation Voltage (I_{SINK} = 2mA)		0.2	0.5		0.2	0.4		0.2	0.4	V
Rise Time (R_L = 4.7kΩ)		100			100			100		ns
Fall Time (R_L = 4.7kΩ)		40			40			40		ns
Duty Cycle Adjust	2		98	2		98	2		98	%
Triangle/Sawtooth/Ramp										
Amplitude (R_T = 100kΩ)	0.30	0.33		0.30	0.33		0.30	0.33		$\times V_S$
Linearity		0.1			0.05			0.05		%
Output Impedance (I_{OUT} = 5mA)		200			200			200		Ω
Sine-Wave										
Amplitude (R_S = 100kΩ)		0.2	0.22		0.2	0.22		0.2	0.22	$\times V_S$
THD (R_S = 1MΩ) Note 4.		0.8	5		0.7	3		0.7	1.5	%
THD Adjusted (Use Fig. 8b)		0.5			0.5			0.5		%

NOTE 1: R_A and R_R currents not included
NOTE 2: V_S = 20V; R_A and R_B = 10kΩ, f ≅ 9kHz; Can be extended to 1000:1 See Figures 13 and 14
NOTE 3: All parameters measured in test circuit given in Fig. 2
NOTE 4: 82 kΩ connected between pins 11 and 12, Triangle Duty Cycle set at 50%. (Use R_A and R_B)
NOTE 5: Derate plastic package at 6.7 mW/°C for ambient temperatures above 50°C
Derate ceramic package at 12.5 mW/°C for ambient temperatures above 100°C
NOTE 6: Over operating temperature range, Fig. 2, pins 7 and 8 connected, V_S = ± 10V. See Fig. 6c for T.C. vs V_S

TEST CONDITIONS (See Fig. 2)

PARAMETER	R_A	R_B	R_L	C_1	SW_1	MEASURE
Supply Current	10kΩ	10kΩ	10kΩ	3.3nF	Closed	Current into Pin 6
Maximum Frequency of Oscillation	1kΩ	1kΩ	4.7kΩ	None	Closed	Frequency at Pin 9
Sweep FM Range (Note 1)	10kΩ	10kΩ	10kΩ	3.3nF	Open	Frequency at Pin 9
Frequency Drift with Temperature	10kΩ	10kΩ	10kΩ	3.3nF	Closed	Frequency at Pin 9
Frequency Drift with Supply Voltage (Note 2)	10kΩ	10kΩ	10kΩ	3.3nF	Closed	Frequency at Pin 9
Output Amplitude: Sine	10kΩ	10kΩ		3.3nF	Closed	Pk–Pk output at Pin 2
Leakage Current (off) Note 3	10kΩ	10kΩ		3.3nF	Closed	Current into Pin 9
(on) Note 3	10kΩ	10kΩ	10kΩ	3.3nF	Closed	Pk–Pk output at Pin 3
Saturation Voltage	10kΩ	10kΩ	10kΩ	3.3nF	Closed	Output (low) at Pin 9
Rise and Fall Times	10kΩ	10kΩ	4.7kΩ	3.3nF	Closed	Waveform at Pin 9
Duty Cycle Adjust: MAX	50kΩ	~1.6kΩ	10kΩ	3.3nF	Closed	Waveform at Pin 9
MIN	~25kΩ	50kΩ	10kΩ	3.3nF	Closed	Waveform at Pin 9
Triangle Waveform Linearity	10kΩ	10kΩ	10kΩ	3.3nF	Closed	Waveform at Pin 3
Total Harmonic Distortion	10kΩ	10kΩ	10kΩ	3.3nF	Closed	Waveform at Pin 2

NOTE 1: The hi and lo frequencies can be obtained by connecting pin 8 to pin 7 (f_{hi}) and then connecting pin 8 to pin 6 (f_{lo}). Otherwise apply Sweep Voltage at pin 8(2/3 VCC + 2V)≤ VSWEEP ≤ VCC where VCC is the total supply voltage. In Fig. 2, Pin 8 should vary between 5.3V and 10V with respect to ground.

NOTE 2: 10V ≤ VCC ≤ 30V, or ±5V ≤ Vs ≤ ±15V.

NOTE 3: Oscillation can be halted by forcing pin 10 to ground.

TEST CIRCUIT

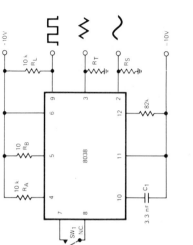

FIGURE 2

DEFINITION OF TERMS:

Supply Current
The current required from the power supply to operate the device, excluding load currents and the currents through R_A and R_B.

Frequency Range
The frequency range at the square wave output through which circuit operation is guaranteed.

Sweep FM Range
The ratio of maximum frequency to minimum frequency which can be obtained by applying a sweep voltage to Pin 8. For correct operation, the sweep voltage should be within the range $(2/3\ V_{CC} + 2V) < V_{sweep} < V_{CC}$.

FM linearity
The percentage deviation from the best-fit straight line on the control voltage versus output frequency curve.

Frequency Drift with Temperature
The change in output frequency as a function of temperature.

Frequency Drift with Supply Voltage
The change in output frequency as a function of supply voltage.

Output Amplitude
The peak-to-peak signal amplitude appearing at the outputs.

Saturation Voltage
The output voltage at the collector of Q_{23} when this transistor is turned on. It is measured for a sink current of 2 mA.

Rise Time and Fall Time
The time required for the square wave output to change from 10% to 90%, or 90% to 10%, of its final value.

Triangle Waveform Linearity
The percentage deviation from the best-fit straight line on the rising and falling triangle waveform.

Total Harmonic Distortion
The total harmonic distortion at the sine-wave output.

279

CHARACTERISTIC CURVES

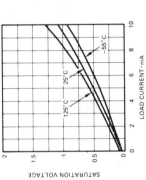

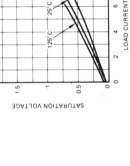

FIGURE 3. PERFORMANCE OF THE SQUARE-WAVE OUTPUT (PIN 9).

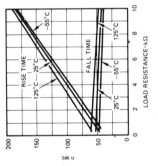

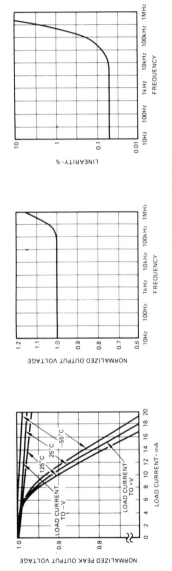

FIGURE 4. PERFORMANCE OF TRIANGLE-WAVE OUTPUT.

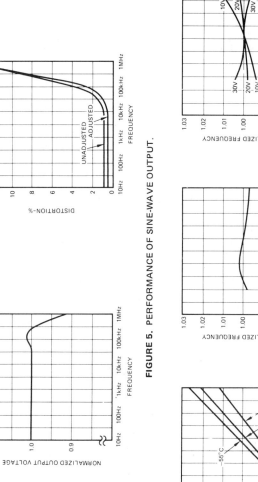

FIGURE 5. PERFORMANCE OF SINE-WAVE OUTPUT.

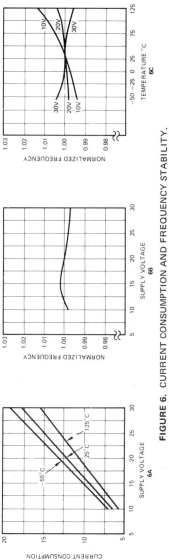

FIGURE 6. CURRENT CONSUMPTION AND FREQUENCY STABILITY.

XR-2206

Monolithic Function Generator *

The XR-2206 is a monolithic function generator integrated circuit capable of producing high quality sine, square, triangle, ramp and pulse waveforms of high stability and accuracy. The output waveforms can be both amplitude and frequency modulated by an external voltage. Frequency of operation can be selected externally over a range of 0.01 Hz to more than 1 MHz.

The XR-2206 is ideally suited for communications, instrumentation, and function generator applications requiring sinusoidal tone, AM, FM or FSK generation. It has a typical drift specification of 20 ppm/°C. The oscillator frequency can be linearly swept over a 2000:1 frequency range with an external control voltage with very little affect on distortion.

As shown in Figure 1, the monolithic circuit is comprised of four functional blocks: a voltage-controlled oscillator (VCO); an analog multiplier and sine-shaper; a unity gain buffer amplifier; and a set of current switches. The internal current switches transfer the oscillator current to any one of the two external timing resistors to produce two discrete frequencies selected by the logic level at the FSK input terminal (pin 9).

FEATURES

Low Sinewave Distortion (THD .5%) –
   insensitive to signal sweep
Excellent Stability (20 ppm/°C, typ)
Wide Sweep Range (2000:1, typ)
Low Supply Sensitivity (0.01%/V, typ)
Linear Amplitude Modulation
Adjustable Duty-Cycle (1% to 99%)
TTL Compatible FSK Controls
Wide Supply Range (10V to 26V)

ABSOLUTE MAXIMUM RATINGS

Power Supply	26V
Power Dissipation	750 mW
Derate above 25°C	5 mW/°C
Total Timing Current	6 mA
Operating Temperature	
Storage Temperature	−65°C to +150°C

*Courtesy of Exar Integrated Systems, Inc.

APPLICATIONS

Waveform Generation
  Sine, Square, Triangle, Ramp
Sweep Generation
AM/FM Generation
FSK and PSK Generation
Voltage-to-Frequency Conversion
Tone Generation
Phase-Locked Loops

AVAILABLE TYPES

Part Number	Package Types (16 Pin DIP)	Operating Temperature Range
XR-2206M	Ceramic	$-55°$C to $+125°$C
XR-2206N	Ceramic	$0°$C to $+75°$C
XR-2206P	Plastic	$0°$C to $+75°$C
XR-2206CN	Ceramic	$0°$C to $+75°$C
XR-2206CP	Plastic	$0°$C to $+75°$C

FUNCTIONAL BLOCK DIAGRAM

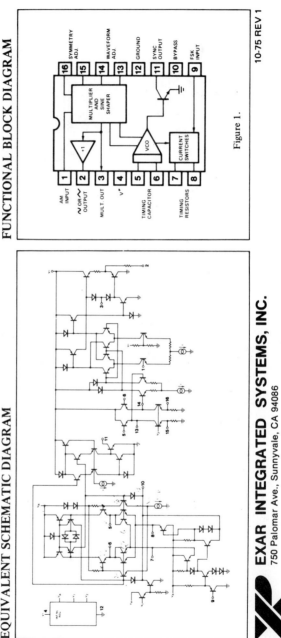

Figure 1.

10-75 REV 1

EQUIVALENT SCHEMATIC DIAGRAM

EXAR INTEGRATED SYSTEMS, INC.
750 Palomar Ave., Sunnyvale, CA 94086

283

ELECTRICAL CHARACTERISTICS

Test Conditions: Test Circuit of Fig. 2, $V^+ = 12V$, $T_A = 25°C$, $C = 0.01\ \mu F$, $R_1 = 100\ K\Omega$, $R_2 = 10\ K\Omega$, $R_3 = 25\ K\Omega$ unless otherwise specified. S_1 open for triangle, closed for sinewave.

CHARACTERISTICS	XR-2206M, XR-2206			XR-2206C			UNITS	CONDITIONS
	MIN.	TYP.	MAX.	MIN.	TYP.	MAX.		
Supply Voltage								
Single Supply	10		26	10		26	V	
Split Supply	±5		±13	±5		±13	V	
Supply Current		12	17		14	20	mA	$R_1 \geq 10\ K\Omega$
Oscillator Section								
Max. Operating Frequency	0.5	1		0.5	1		MHz	$C = 1000$ pF, $R_1 = 1\ K\Omega$
Lowest Practical Frequency		0.01			0.01		Hz	$C = 50\ \mu F$, $R_1 = 2\ M\Omega$
Frequency Accuracy		±1	±4		±2		% of f_o	$f_o = 1/R_1C$
Temperature Stability		±10	±50		±20		ppm/°C	$0°C \leq T_A \leq 75°C$, $R_1 = R_2 = 20\ K\Omega$
Supply Sensitivity		0.01	0.1		0.01		%	$V_{LOW} = 10V$, $V_{HIGH} = 20V$, $R_1 = R_2 = 20\ K\Omega$
Sweep Range	1000:1	2000:1			2000:1		$f_H = f_L$	f_H @ $R_1 = 1\ K\Omega$, f_L @ $R_1 = 2\ M\Omega$
Sweep Linearity								
10:1 Sweep		2			2		%	$f_L = 1$ kHz, $f_H = 10$ kHz
1000:1 Sweep		8			8		%	$f_L = 100$ Hz, $f_H = 100$ kHz
FM Distortion		0.1			0.1		%	±10% Deviation
Recommended Timing Components								
Timing Capacitor: C	0.001		100	0.001		100	μF	See Figure 5
Timing Resistors: R_1 & R_2	1		2000	1		2000	$K\Omega$	
Triangle/Sinewave Output								
Triangle Output	40	160	80		160		mV/KΩ	See Note 1, Fig4, Fig. 2 S_1 Open
Sinewave Output		60			60		mV/KΩ	Fig. 2 S_1 Closed
Max. Output Swing		6			6		Vpp	
Output Impedance		600			600		Ω	
Triangle Linearity		1			1		%	
Amplitude Stability		0.5			0.5		dB	For 1000:1 Sweep

Parameters							UNIT	CONDITIONS
Sinewave Distortion								
Without Adjustment		2.5			2.5		%	$R_1 = 30\ K\Omega$
With Adjustment		0.4	1.0		0.5	1.5	%	See Figure 11. / See Figure 12
Amplitude Modulation								
Input Impedance	50	100			100		$K\Omega$	
Modulation Range		100			100		%	
Carrier Suppression		55			55		dB	
Linearity		2			2		%	For 95% modulation
Square Wave Output								
Amplitude		12			12		Vpp	Measured at Pin 11
Rise Time		250			250		nsec	$C_L = 10$ pF
Fall Time		50			50		nsec	$C_L = 10$ pF
Saturation Voltage		0.2	0.4		0.2	0.4	V	$I_L = 2$ mA
Leakage Current		0.1	20				μA	$V_{11} = 12$V
FSK Keying Level (Pin 9)	0.8	1.4	2.4	0.8	1.4	2.5	V	See Section on Circuit Controls
Reference Bypass Voltage	2.9	3.1	3.3	2.9	3	3.5	V	Measured at Pin 10.

Note 1: Output Amplitude is inversely proportional to the resistance R_3 on Pin 3. See Figure 3

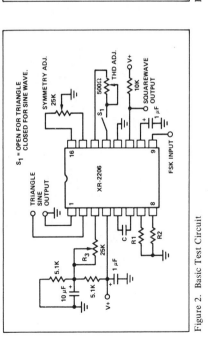

Figure 2. Basic Test Circuit

Figure 3. Output Amplitude as a Function of Resistor R_3 at Pin 3.

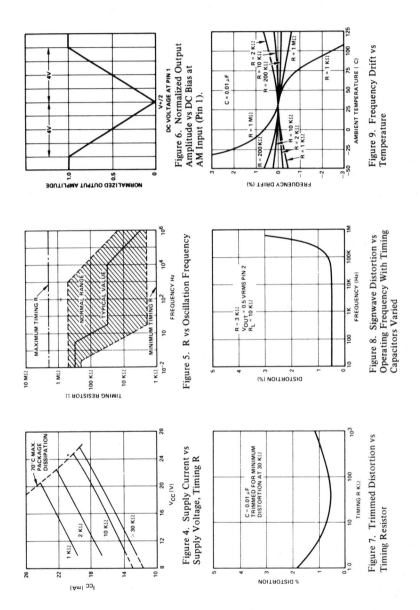

Figure 4. Supply Current vs Supply Voltage, Timing R

Figure 5. R vs Oscillation Frequency

Figure 6. Normalized Output Amplitude vs DC Bias at AM Input (Pin 1).

Figure 7. Trimmed Distortion vs Timing Resistor

Figure 8. Signwave Distortion vs Operating Frequency With Timing Capacitors Varied

Figure 9. Frequency Drift vs Temperature

286

DESCRIPTION OF CIRCUIT CONTROLS

FREQUENCY OF OPERATION:

The frequency of oscillation, f_o, is determined by the external timing capacitor C across pins 5 and 6, and by the timing resistor R connected to either pin 7 or pin 8. The frequency is given as

$$f_o = \frac{1}{RC} \quad Hz$$

and can be adjusted by varying either R or C. The recommended values of R for a given frequency range are shown in Figure 5. Temperature stability is optimum for $4\ K\Omega < R < 200\ K\Omega$. Recommended values of C are from 1000 pF to 100 μF.

FREQUENCY SWEEP AND MODULATION

Frequency of oscillation is proportional to the *total* timing current I_T drawn from pin 7 or 8

$$f = \frac{320 I_T\ (mA)}{C\ (\mu F)} \quad Hz$$

Timing terminals (pins 7 or 8) are low impedance points and are internally biased at +3V, with respect to pin 12. Frequency varies linearly with I_T over a wide range of current values, from 1 μA to 3 mA. The frequency can be controlled by applying a control voltage, V_C, to the activated timing pin as shown in Figure 10. The frequency of oscillation is related to V_C as:

$$f = \frac{1}{RC}\left[1 + \frac{R}{R_C}\left(1 - \frac{V_C}{3} \right) \right]\ Hz$$

where V_C is in volts. The voltage-to-frequency conversion gain, K, is given as:

$$K = \partial f / \partial V_C = -\frac{0.32}{R_C C} \quad Hz/V$$

NOTE: For safe operation of the circuit I_T should be limited to ≤ 3 mA.

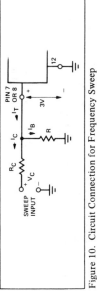

Figure 10. Circuit Connection for Frequency Sweep

OUTPUT CHARACTERISTICS:

Output Amplitude: Maximum output amplitude is inversely proportional to external resistor R_3 connected to Pin 3 (See Fig. 3). For sinewave output, amplitude is approximately 60 mV peak per $K\Omega$ of R_3; for triangle, the peak amplitude is approximately 160 mV peak per $K\Omega$ of R_3. Thus, for example, $R_3 = 50\ K\Omega$ would produce approximately ±3V sinusoidal output amplitude.

Amplitude Modulation: Output amplitude can be modulated by applying a dc bias and a modulating signal to Pin 1. The internal impedance at Pin 1 is approximately 200 $K\Omega$. Output amplitude varies linearly with the applied voltage at Pin 1, for values of dc bias at this pin, within ±4 volts of $V^+/2$ as shown in Fig. 6. As this bias level approaches $V^+/2$, the phase of the output signal is reversed; and the amplitude goes through zero. This property is suitable for phase-shift keying and suppressed-carrier AM generation. Total dynamic range of amplitude modulation is approximately 55 dB.

Note: AM control must be used in conjunction with a well-regulated supply since the output amplitude now becomes a function of V^+.

FREQUENCY-SHIFT KEYING

The XR-2206 can be operated with two separate timing resistors, R_1 and R_2, connected to the timing pins 7 and 8, respectively, as shown in Figure 13. Depending on the polarity of the logic signal at pin 9, either one or the other of these timing

287

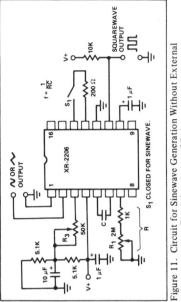

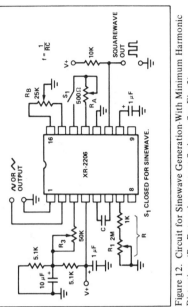

Figure 11. Circuit for Sinewave Generation Without External Adjustment. (See Fig. 3 for choice of R_3)

Figure 12. Circuit for Sinewave Generation-With Minimum Harmonic Distortion. (R_3 Determines output Swing – See Fig. 3)

resistors is activated. If pin 9 is open-circuited or connected to a bias voltage $\geq 2V$, only R_1 is active. Similarly, if the voltage level at pin 9 is $\leq 1V$, only R_2 is activated. Thus, the output frequency can be keyed between two levels, f_1 and f_2 as:

$$f_1 = 1/R_1 C \text{ and } f_2 = 1/R_2 C$$

For split-supply operation, the keying voltage at pin 9 is referenced to V^-

OUTPUT DC LEVEL CONTROL

The dc level at the output (pin 2) is approximately the same as the dc bias at pin 3. In Figures 11, 12 and 13, pin 3 is biased mid-way between V^+ and ground, to give an output dc level of $\approx V^+/2$.

APPLICATIONS INFORMATION

SINEWAVE GENERATION

A) Without External Adjustment

Figure 11 shows the circuit connection for generating a sinusoidal output from the XR-2206. The potentiometer R_1 at pin 7 provides the desired frequency tuning. The maximum output swing is greater than $V^+/2$ and the typical distortion (THD) is < 2.5%. If lower sinewave distortion is desired, additional adjustments can be provided as described in the following section.

The circuit of Figure 11 can be converted to split supply operation simply by replacing all ground connections with V^-. For split supply operation, R_3 can be directly connected to ground.

B) With External Adjustment

The harmonic content of sinusoidal output can be reduced to $\approx 0.5\%$ by additional adjustments as shown in Figure 12 The potentiometer R_A adjusts the sine-shaping resistor;

288

and R_B provides the fine-adjustment for the waveform symmetry. The adjustment procedure is as follows:

1. Set R_B at mid-point and adjust R_A for minimum distortion.

2. With R_A set as above, adjust R_B to further reduce distortion.

TRIANGLE WAVE GENERATION

The circuits of Figures 11 and 12 can be converted to triangle wave generation by simply open circuiting pins 13 and 14 (i.e., S_1 open). Amplitude of the triangle is approximately twice the sinewave output.

FSK GENERATION

Figure 13 shows the circuit connection for sinusoidal FSK signal generation. Mark and space frequencies can be independently adjusted by the choice of timing resistors R_1 and R_2; and the output is phase-continuous during transitions. The keying signal is applied to pin 9. The circuit can be converted to split-supply operation by simply replacing ground with V^-.

PULSE AND RAMP GENERATION

Figure 14 shows the circuit for pulse and ramp waveform generation. In this mode of operation, the FSK keying terminal (pin 9) is shorted to the square-wave output (pin 11); and the circuit automatically frequency-shift keys itself between two separate frequencies during the positive and negative going output waveforms. The pulse-width and the duty cycle can be adjusted from 1% to 99% by the choice of R_1 and R_2. The values of R_1 and R_2 should be in the range of 1 KΩ to 2 MΩ.

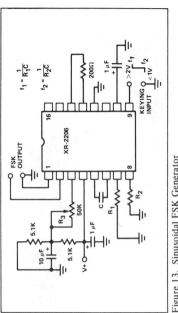

$$f_1 = \frac{1}{R_1 C}$$

$$f_2 = \frac{1}{R_2 C}$$

Figure 13. Sinusoidal FSK Generator

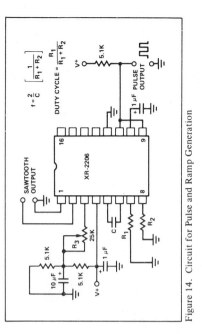

$$f = \frac{2}{C}\left[\frac{1}{R_1 + R_2}\right]$$

$$\text{DUTY CYCLE} = \frac{R_1}{R_1 + R_2}$$

Figure 14. Circuit for Pulse and Ramp Generation

289

IERC

Heat Dissipators for
Case-Mounted Metal Medium
and High-Power Semiconductors *

UP-10

LB66B1

LB66B2

UP

UP2

UP1

UP3

LA

LA

HP3

HP1

*Courtesy of International Electronic Research Corporation

X-75 TO-8 DO-4 TO-15 TO-6 TO-66 TO-3
                          TO-36

Patent No. 3,212,569

Representative semiconductors which can be used with the dissipators shown above.

Unique Staggered Finger Configuration is Smaller and Lighter Than Finned Extrusions; More Efficient

Standard models dissipate from 1 to 80 watts in natural convection; up to 150 watts in forced air environments.

IERC's Power Series heat sinks/dissipators provide efficient, low cost cooling for a wide range of medium and high power dissipation requirements. The HP3, for example, has 30.2 square inches of surface, requires only 9 cubic inches of space, and weighs only 1.5 ounces. But it dissipates as much heat as many conventional finned extrusions which have ⅓ more volume and 3 times the weight.

The secret is in the design

The high volume-efficiency ratio of these dissipators is due to IERC's patented design. In still air the separate, multiple, staggered fingers dissipate, by radiation and convection, directly to the ambient — not to another finger surface. Conversely, extrusion fins radiate to each other. And, the free movement of convection currents is hampered by their being confined in the deep cavities between the fins.

Forced air increases efficiency

In forced air the IERC Power Series dissipators are even more efficient. The staggered fingers increase turbulence, di-

recting air completely around each finger for maximum dissipation. But, with finned extrusions, the forced air begins to leave the surfaces immediately. By the time it is part way down the extrusion it is hitting only the top edges of the fins, resulting in minimal dissipation.

Also, extrusions are directional, and reach a maximum efficiency at a specified air velocity dictated by their fin size and spacing. IERC Power Series dissipators can be mounted in any vertical or horizontal mode. They give the same high efficiency with forced air from any direction and over a broad range of velocities.

For dissipators with IC hole patterns see Bulletins 161, 162, 163, 169, and 175

291

LB SERIES

In natural convection, the LB's efficient staggered finger design dissipates 9 watts with case temperature held to less than 150°C. In forced air (1,000 fpm) the LB dissipates 15 watts with case temperature maintained at 90°C. Aluminum construction; requires only 1.7 sq. in. of mounting area.

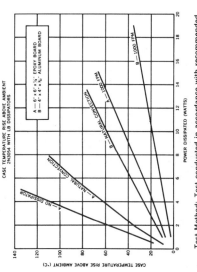

CASE TEMPERATURE RISE ABOVE AMBIENT
2N3054 WITH LB DISSIPATORS

A — 6″ x 6″ x ½″ EPOXY BOARD
B — 4″ x 4″ x ⅛″ ALUMINUM BOARD

Test Method: Test conducted in accordance with recommended procedures in EIA Components Bulletin No. 5.

LB66B1U

For complete dimensions and tolerances request drawing.

Ordering Information

	IERC PART NO.				
Unplated	Com'l. Black Anodize	Mil. Black Anodize	Insulube® 448	Dimension "C"	Semiconductor
LB66B1U	LB66B1CB	LB66B1B	LB66B1	½″ H	1 TO-66
LB66B2U	LB66B2CB	LB66B2B	LB66B2	⅜″ H	1 TO-66

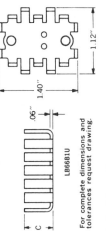

LA SERIES

Power increases of two to six times over bare case semiconductor operation may be realized with these low cost "limited area" dissipators for TO-3 and TO-66 transistors and ICs and plastic case packages such as TO-126 and C-106. For example, the LAD66A4U in natural convection will allow a 2N3054 TO-66 transistor to dissipate over 6 watts with a 90°C case rise. This compares with less than 3 watts for a bare case device in free air. Both LA models are made of aluminum in 3 different heights and feature IERC's patented staggered finger design. Various finishes can be provided including IERC's exclusive Insulube® 448 for positive 500 volt dielectric protection of hot case semiconductors.

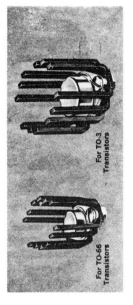

For TO-3 Transistors

For TO-66 Transistors

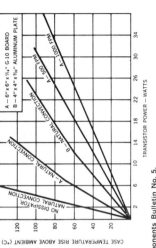

2N3055 (TO-3) WITH LAT03B5CB

A — 6" x 6" x ¹⁄₁₆" G-10 BOARD
B — 4" x 4" x ³⁄₃₂" ALUMINUM PLATE

CASE TEMPERATURE RISE ABOVE AMBIENT (°C)
TRANSISTOR POWER — WATTS

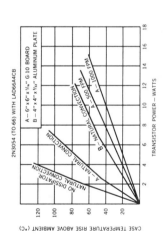

2N3054 (TO-66) WITH LAD66A4CB

A — 6" x 6" x ¹⁄₁₆" G-10 BOARD
B — 4" x 4" x ³⁄₃₂" ALUMINUM PLATE

CASE TEMPERATURE RISE ABOVE AMBIENT (°C)
TRANSISTOR POWER — WATTS

LAD66

LAT03

For complete dimensions and tolerances request drawing.

Test Method: Test conducted in accordance with recommended procedures in EIA Components Bulletin No. 5.

Ordering Information

IERC PART NO.						
Unplated	Com'l. Black Anodize	Mil. Black Anodize	Insulube® 448	Dimension "A"	Semiconductor	
LAD66A1U	LAD66A1CB	LAD66A1B	LAD66A1	¼"	TO-66	
LAD66A2U	LAD66A2CB	LAD66A2B	LAD66A2	½"	TO-66	
LAD66A3U	LAD66A3CB	LAD66A3B	LAD66A3	¾"	TO-66	
LAD66A4U	LAD66A4CB	LAD66A4B	LAD66A4	1"	TO-66	
LAT03B3U	LAT03B3CB	LAT03B3B	LAT03B3	¾"	TO-3	
LAT03B4U	LAT03B4CB	LAT03B4B	LAT03B4	1"	TO-3	
LAT03B5U	LAT03B5CB	LAT03B5B	LAT03B5	1¼"	TO-3	

293

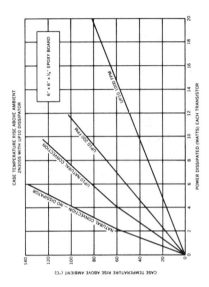

CASE TEMPERATURE RISE ABOVE AMBIENT
2N3055 WITH UP10 DISSIPATOR

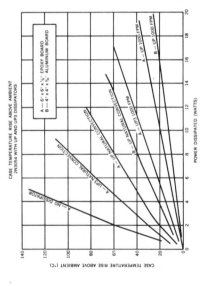

CASE TEMPERATURE RISE ABOVE AMBIENT
2N3054 WITH UP AND UP3 DISSIPATORS

UP SERIES

Staggered-finger design and aluminum construction gives excellent dissipation on medium power applications. For example, a silicon power transistor mounted in a UP will dissipate 14 watts in natural convection with a maximum case rise of 100°C. In forced air (1,000 fpm), the UP dissipates 50 watts with the same case temperature rise. Requires only 3 cu. in. of space; weighs less than 1 ounce.

Compact UP10 accommodates TO3 pairs — Two TO3's mounted in a UP10 will dissipate 20 watts in still air with a case temperature rise of less than 130°C — nearly twice as much power as the same TO3's without the UP10. The compact size of the UP10 makes it ideal for high density applications where boards are stacked on ½" centers.

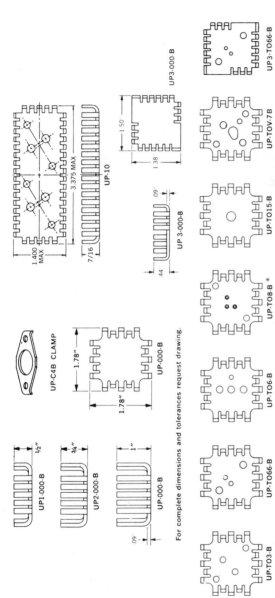

For complete dimensions and tolerances request drawing.

Hole Patterns

UP Dissipators incorporate standard hole configurations for many power transistors and diodes. However, the basic UP-000-B can be supplied with hole configurations to your specific requirements. The drawings above will enable you to relate a particular JEDEC outline or popular transistor package to the correct UP series dissipator. For more detailed dimensions ask for IERC drawings.

For a specific transistor hole pattern, replace the "000" in the part number with the transistor designation shown above. For example: UP-TO3-U. For alternate hole patterns in the UP10 contact the factory.

For a universal hole pattern, replace the "000" with "TOV-7". For example: UP-TOV-7U. Hole pattern accommodates TO-3, TO-6, TO-36 and other stud-mounted transistors up to 1/4" diameter.

*UP Series dissipators will accommodate TO-8 case size semiconductors when a UP-C4U, UP-C4CB, UP-C4B or UP-C4 clamp is used.

Ordering Information

| Unplated | IERC PART NO. | | | Semiconductor |
	Com'l. Black Anodize	Mil. Black Anodize	Insulube® 448	
UP-000-U	UP-000-CB	UP-000-B	UP-000	Undrilled
UP1-000-U	UP1-000-CB	UP1-000-B	UP1-000	Undrilled
UP2-000-U	UP2-000-CB	UP2-000-B	UP2-000	Undrilled
UP3-000-U	UP3-000-CB	UP3-000-B	UP3-000	Undrilled
UP10-TO3-2U	UP10-TO3-2CB	UP10-TO3-2B	UP10-TO3-2	2 TO-3's

HP SERIES — Single semiconductor mounting

The heat dissipating performance of the HP's patented staggered-finger design and aluminum construction is comparable to conventional finned extrusions up to 3 times heavier and 30% larger. For example, the HP1 dissipates 23 watts in natural convection with a maximum case temperature rise of 100°C. The HP3 dissipates 28 watts with the same case temperature rise. In a forced air mode (1,000 fpm), the HP1 dissipates 65 watts; the HP3, 74 watts, with a 100°C case temperature rise. When nested together in a forced air mode (1,000 fpm), more than 100 watts can be dissipated at the same 100°C case temperature rise.

The graphs below show the case temperatures of 2N1899 transistors with HP Series dissipators mounted on an epoxy board in a 25°C ambient.

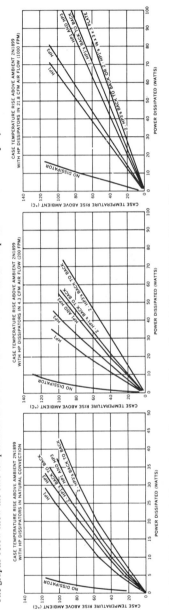

Test Method: Test conducted in accordance with recommended procedures in EIA Components Bulletin No. 5.

296

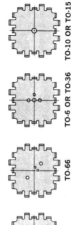

HP1-OOO-B

.90" .09" 2½" (TYP.)

HP3-OOO-B

1.0" .09" 3¼" (TYP.)

For complete dimensions and tolerances request drawing.

Non-Standard Heights Both HP1 and HP3 parts can be furnished as non-standards with either ½" or ¾" heights. For ½" height add -6 to the part number (Example: HP1-TO3-6U). For ¾" add -3 at the end of the part number (Example: HP1-TO3-3U).

Ordering Information

	IERC PART NO.			
Unplated	Com'l. Black Anodize	Mil. Black Anodize	Insulube® 448	Semiconductor
HP1-000-U	HP1-000-CB	HP1-000-B	HP1-000	Undrilled
HP3-000-U	HP3-000-CB	HP3-000-B	HP3-000	Undrilled

For a specific transistor hole pattern, replace the "000" with the transistor designation. For example: HP-TO3-U.

For a universal hole pattern, replace the "000" with "TOV-2". For example: HP1-TOV-2U. Hole pattern accommodates TO-3, TO-6, TO-36 and other stud-mounted transistors up to ¼" diameter.

Hole Patterns

TO-3   TO-66   TO-6 OR TO-36   TO-10 OR TO-15

Other Hole Patterns HP dissipators with standard hole patterns are readily available from stocking IERC Authorized Distributors. Other patterns available from factory on request at small additional cost.

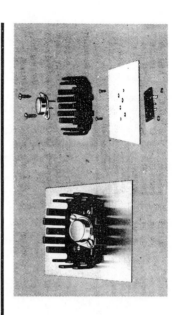

Panel mounted models allow easy transistor removal

HP1 and HP3 models can be furnished with a new standard hole pattern which greatly simplifies equipment assembly and provides for plug-in transistor removal and replacement. The dissipators feature a TO-3 semiconductor hole pattern with clearance holes for socket hardware, enabling removal of the dissipator and transistor from the chassis mounting surface without disassembly of the transistor socket or its hardware.

Ordering Information

	IERC PART NO. — PANEL MOUNTED			
Unplated	Com'l. Black Anodize	Mil. Black Anodize	Insulube® 448	Semiconductor
HP1-TO3-44U	HP1-TO3-44CB	HP1-TO3-44B	HP1-TO3-44	TO-3
HP3-TO3-44U	HP1-TO3-44CB	HP1-TO3-44B	HP1-TO3-44	TO-3

HP SERIES — Multiple semiconductor mounting

The standard HP1 is only 2½" square and .90" high but will accommodate three TO-66 case size semiconductors. The HP3 is 3⅛" square and 1" high, yet accommodates four TO-3 or TO-66 semiconductors. Use of these compact, lightweight dissipators eliminates heavy, bulky finned extrusions.

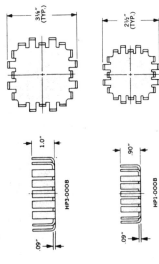

HP3-000B

HP1-000B

For complete dimensions and tolerances request drawing.

Ordering Information

| Unplated | IERC PART NO. | | | | Semi-conductor |
	Com'l. Black Anodize	Mil. Black Anodize	Insulube® 448	
HP1-T066-4U	HP1-T066-4CB	HP1-T066-4B	HP1-T066-4	2 TO-66
HP1-T066-8U	HP1-T066-8CB	HP1-T066-8B	HP1-T066-8	3 TO-66
HP3-T03-4U	HP3-T03-4CB	HP3-T03-4B	HP3-T03-4	2 TO-3
HP3-T03-8U	HP3-T03-8CB	HP3-T03-8B	HP3-T03-8	3 TO-3
HP3-T03-11U	HP3-T03-11CB	HP3-T03-11B	HP3-T03-11	4 TO-3
HP3-T066-4U	HP3-T066-4CB	HP3-T066-4B	HP3-T066-4	2 TO-66
HP3-T066-8U	HP3-T066-8CB	HP3-T066-8B	HP3-T066-8	3 TO-66
HP3-T066-11U	HP3-T066-11CB	HP3-T066-11B	HP3-T066-11	4 TO-66

Hole patterns for multiple mounting of other high power semiconductors can be provided at small additional cost.

Hole Patterns

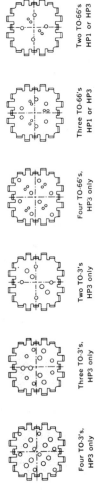

Two TO-66's
HP1 or HP3

Three TO-66's
HP1 or HP3

Four TO-66's,
HP3 only

Two TO-3's
HP3 only

Three TO-3's,
HP3 only

Four TO-3's,
HP3 only

Dissipation curves for multiple mounting of TO-3 and TO-66 transistors

The following is thermal performance data for 2, 3, and 4 TO-3 transistors mounted in an HP3 series dissipator, and 2, 3 and 4 TO-66 transistors mounted in HP1 and HP3 dissipators.

In addition to allowing high power dissipation from multiple transistors, Power Series dissipators also allow thermal matching of transistors for applications such as the Darlington amplifier. In all tests case temperatures of all transistors were matched within 1° or 2°C.

The transistors used in these tests were silicon TO-3 type 2N3055 and silicon TO-66 type 2N3054; however, the curves are applicable to any device with the same or similar thermal resistance junction to case.

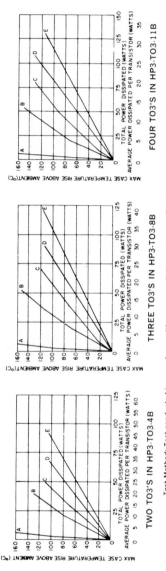

TWO TO3'S IN HP3-TO3-4B

THREE TO3'S IN HP3-TO3-8B

FOUR TO3'S IN HP3-TO3-11B

Test Method: Test conducted in accordance with recommended procedures in EIA Components Bulletin No. 5.

Code letters for TO-3 curves

A   Transistor mounted to a G-10 board — no dissipator — natural convection
B   Transistor in HP3 dissipator mounted to a G-10 board — natural convection
C   Transistor in HP3 dissipator mounted to a 6 x 6 x 3/32 aluminum plate — natural convection
D   Transistor in HP3 dissipator on a G-10 board, in forced convection, 500 fpm
E   Transistor in HP3 dissipator on a G-10 board, in forced convection, 1000 fpm

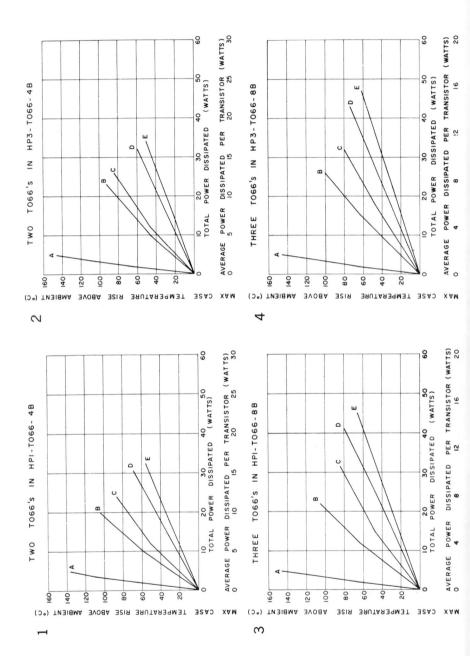

300

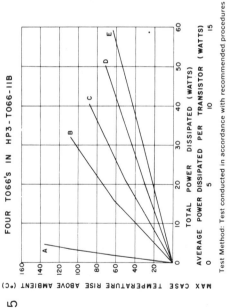

FOUR TO66's IN HP3-TO66-IIB

MAX CASE TEMPERATURE RISE ABOVE AMBIENT (°C)

TOTAL POWER DISSIPATED (WATTS)

AVERAGE POWER DISSIPATED PER TRANSISTOR (WATTS)

5

Code letters for TO-66 curves

A Transistor mounted to a G-10 board, no dissipator, in natural convection.

B Transistor in dissipator mounted to a G-10 board in natural convection.

C Dissipator mounted to an aluminum plate in natural convection. On curves 1 and 2, it is a 4 x 4 x 3/32 aluminum plate. On curves 3, 4, and 5 it is a 6 x 6 x 3/32 aluminum plate.

D Transistor mounted in dissipator on a G-10 board, in forced convection, 500 fpm.

E Transistor mounted in dissipator on a G-10 board, in forced convection, 1000 fpm.

Test Method: Test conducted in accordance with recommended procedures in EIA Components Bulletin No. 5.

INTERNATIONAL ELECTRONIC RESEARCH CORPORATION
a subsidiary of DYNAMICS CORPORATION OF AMERICA
135 West Magnolia Blvd., Burbank, California 91502   213/849-2481

BULLETIN 164

Index